CAMBRIDGE LIBRARY COLLECTION

Books of enduring scholarly value

Botany and Horticulture

Until the nineteenth century, the investigation of natural phenomena, plants and animals was considered either the preserve of elite scholars or a pastime for the leisured upper classes. As increasing academic rigour and systematisation was brought to the study of 'natural history', its subdisciplines were adopted into university curricula, and learned societies (such as the Royal Horticultural Society, founded in 1804) were established to support research in these areas. A related development was strong enthusiasm for exotic garden plants, which resulted in plant collecting expeditions to every corner of the globe, sometimes with tragic consequences. This series includes accounts of some of those expeditions, detailed reference works on the flora of different regions, and practical advice for amateur and professional gardeners.

Ladies' Botany

The horticulturalist John Lindley (1799–1865) worked for Sir Joseph Banks, and was later instrumental in saving the Royal Horticultural Society from financial disaster. He was a prolific author of works for gardening practitioners but also for a non-specialist readership, and many of his books have been reissued in this series. The first volume of this two-volume work was published in 1834, and the second in 1837. At a time when botany was regarded as the only science suitable for study by women and girls, Lindley felt that there was a lack of books for 'those who would become acquainted with Botany as an amusement and a relaxation', and attempted to meet this need. In the second volume of 'this little work', Lindley continues to introduce new 'tribes' of plants, including exotica such as mangoes and Venus's fly-traps, to his lady correspondent and her children.

Cambridge University Press has long been a pioneer in the reissuing of out-of-print titles from its own backlist, producing digital reprints of books that are still sought after by scholars and students but could not be reprinted economically using traditional technology. The Cambridge Library Collection extends this activity to a wider range of books which are still of importance to researchers and professionals, either for the source material they contain, or as landmarks in the history of their academic discipline.

Drawing from the world-renowned collections in the Cambridge University Library and other partner libraries, and guided by the advice of experts in each subject area, Cambridge University Press is using state-of-the-art scanning machines in its own Printing House to capture the content of each book selected for inclusion. The files are processed to give a consistently clear, crisp image, and the books finished to the high quality standard for which the Press is recognised around the world. The latest print-on-demand technology ensures that the books will remain available indefinitely, and that orders for single or multiple copies can quickly be supplied.

The Cambridge Library Collection brings back to life books of enduring scholarly value (including out-of-copyright works originally issued by other publishers) across a wide range of disciplines in the humanities and social sciences and in science and technology.

Ladies' Botany

Or, a Familiar Introduction to the Study of the Natural System of Botany

VOLUME 2

JOHN LINDLEY

CAMBRIDGE
UNIVERSITY PRESS

University Printing House, Cambridge, CB2 8BS, United Kingdom

Cambridge University Press is part of the University of Cambridge.
It furthers the University's mission by disseminating knowledge in the pursuit of education, learning and research at the highest international levels of excellence.

www.cambridge.org
Information on this title: www.cambridge.org/9781108076562

This edition first published 1837
This digitally printed version 2015

ISBN 978-1-108-07656-2 Paperback

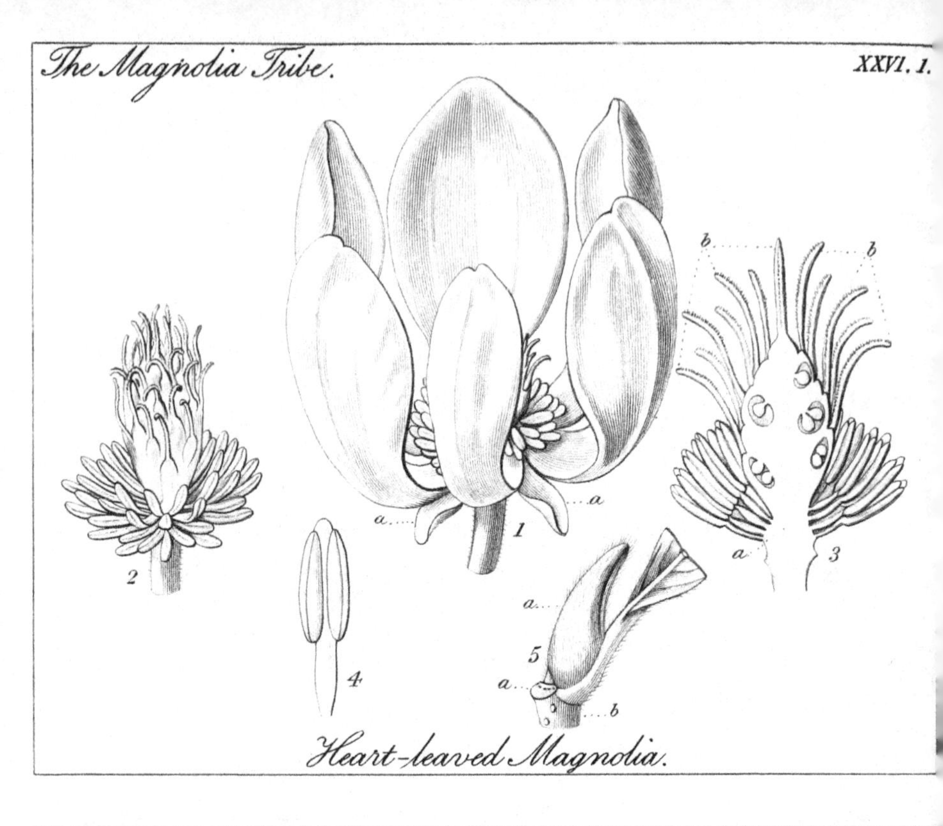

The Magnolia Tribe.
XXVI. 1.
b
b
a
a
1
2
3
a
4
a
5
a
b
Heart-leaved Magnolia.

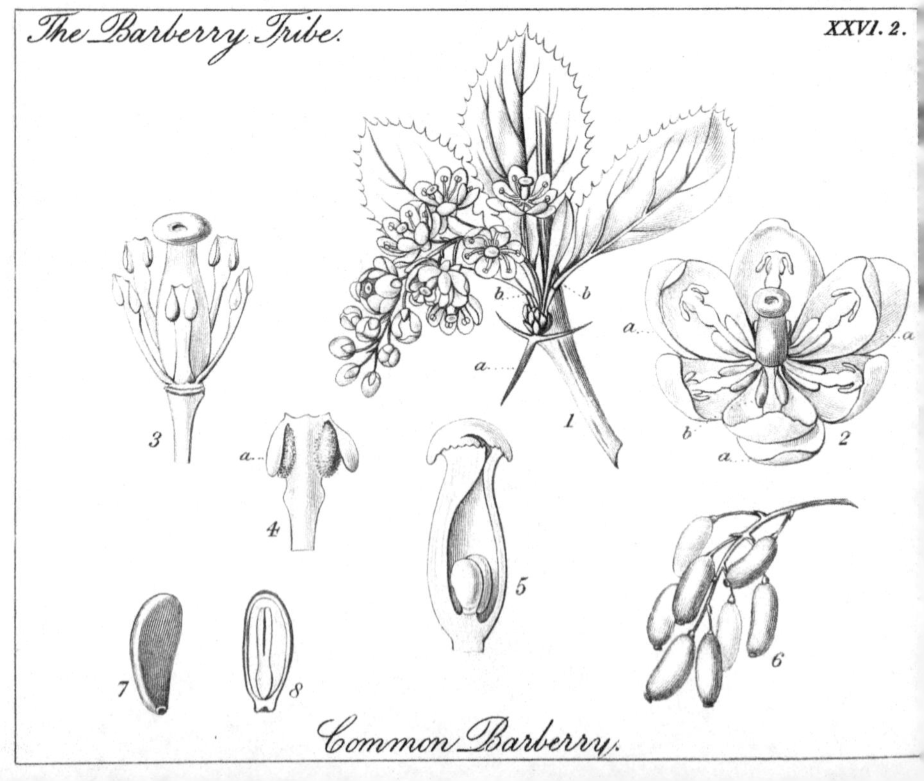

The Barberry Tribe.
XXVI. 2.
b
b
a
a
a
a
b
a
1
2
3
4
5
6
7
8
Common Barberry.

LADIES' BOTANY:

OR

A FAMILIAR INTRODUCTION

To the Study

OF THE

NATURAL SYSTEM OF BOTANY.

BY

JOHN LINDLEY, Ph.D. F.R.S.

ETC. ETC. ETC.

PROFESSOR OF BOTANY IN UNIVERSITY COLLEGE, LONDON.

I boast no song in magic wonders rife,
But yet, oh Nature! is there nought to prize,
Familiar in thy bosom scenes of life?
And dwells in daylight truth's salubrious skies,
No form with which the soul may sympathize?

Campbell.

VOL. II.

LONDON:

JAMES RIDGWAY AND SONS, PICCADILLY.

1837.

LOCKE has two sentences which, with little alteration, express in the best possible manner what I would write upon this occasion.

Nature, he says, commonly lodges her treasure and jewels in rocky ground. If the matter be knotty, and the sense lies deep, the mind must stop and buckle to it, and stick upon it with labour and thought, and close contemplation, and not leave it until it has mastered the difficulty and got possession of truth. And again—

God has made the organic world harmonious and beautiful without us; but it will never come into our heads all at once; we must bring it home piece-meal, and then set it up by our own industry, or else we shall have nothing but darkness and a chaos within, whatever order and light there be in things without us.

The reader will, I am sure, forgive me if I intro duce the second volume of this little work without further preface.

TABLE OF CONTENTS.

LETTERS ON BOTANY.

LETTER XXVI.

PRELIMINARY—THE MAGNOLIA TRIBE—THE BARBERRY TRIBE.

(*Plate XXVI.*)

At the close of our Botanical correspondence two years ago, I had given you an unfinished, but rather extensive, sketch, of the structure and classification of a considerable number of interesting plants; and I then supposed that, for all elementary purposes, I had already occupied so much of your time as to have run the risk of wearying rather than amusing you. The sketch, indeed, was far from comprehending all the beautiful objects by which the admiration of a lover of flowers is excited, nor did it include a complete view of even the most common species that are yielded by our own woods and fields and hills; but it furnished you with a plan of study, it taught you the right manner of exercising your powers of observation, it explained many of the more important facts connected with the organization of the Vegetable World, and it was calculated to place you in a position from which you might proceed as much further in the pursuit of this pleasing science as taste or opportunity might lead you. I had no expectation that my letters would form even an epitome of the most common facts. They were merely intended as an experiment upon the possibility of conveying strictly scientific knowledge in a simple and amusing form, and of showing that Botany is by no means that dry, difficult, repulsive subject, which it may well appear to those who only know it through the uninviting medium of systematic works. I thought it practicable, without at all deserting science, to divest her of the severe, forbidding features that she puts on when dressed in the starched, old-fashioned, matter-of-fact costume of the schools, and to shew that it is in her wild and unsophisticated state that she shines forth in all her smiles and loveliness, when her flowers are newly gathered, their colours fresh, and their fragrance unimpaired, and not when every thing is dry and withered, and formally labelled with the Greek and Latin names of science. I was, moreover, anxious that the endless variety of beautiful objects which the Vegetable world so prodigally strews before our path should, with those who from their habits of life and their gentler feelings are the most sensible to the charms of nature, become something beyond a vague sentiment of undefined admiration. The love for flowers is a holy feeling, inseparable from our very nature ; it exists alike in savage and civilized society ; it speaks with the same powerful voice to the great and wealthy and to the poor and lowly ; it grows up and flourishes with our innocence, and it only perishes with the best and truest feelings of humanity.

——————— O Father, Lord!
The All-Beneficent! I bless thy name,
That thou hast mantled the green earth with flowers,
Linking our hearts to nature! By the love
Of their wild blossoms, our young footsteps first
Into her deep recesses are beguiled,
Her minster cells; dark glen and forest bower,
Where, thrilling with its earliest sense of Thee,
Amidst the low religious whisperings
And shivering leaf-sounds of the solitude,
The spirit wakes to worship, and is made
Thy living temple. By the breath of flowers
Thou callest us, from city throngs and cares,
Back to the woods, the birds, the mountain streams,
That sing of Thee! back to free childhood's heart,
Fresh with the dews of tenderness!—Thou bidd'st
The lilies of the field with placid smile
Reprove man's feverish strivings, and infuse
Through his warm soul a more unworldly life,
With their soft gentle breath. Thou hast not left
His purer nature, with its fine desires,
Uncared for in this universe of thine!
The glowing rose attests it, the beloved
Of poet hearts, touched by their fervent dreams
With spiritual light, and made a source
Of heaven-ascending thoughts. E'en to faint age
Thou lend'st the vernal bliss:—the old man's eye
Falls on the kindling blossoms, and his soul
Remembers youth and love, and hopefully
Turns unto Thee, who call'st earth's buried germs
From dust to splendour; as the mortal seed
Shall, at thy summons, from the grave spring up
To put on glory, to be girt with power,

And filled with immortality. Receive
Thanks, blessings, love, for these, thy lavish boons,
And, most of all, their heavenward influences,
O Thou that gav'st us flowers!

"La vue d'une fleur," says Madame Roland, "caresse mon imagination et flatte mes sens à un point inexprimable—elle reveille avec volupté le sentiment de mon existence—sous le tranquille abri du toit paternel, j'etais heureuse dès l'enfance avec des fleurs et des livres—dans l'étroite enceinte d'une prison, au milieu des fers imposés par la tyrannie la plus révoltante, j'oublie l'injustice des hommes, leurs sottises et mes maux, avec des livres et des fleurs."

How much stronger and more permanent an influence must those feelings exercise upon our nature when the lovely objects that give rise to them are known by something beyond a name, or a favourite colour, or a delightful fragrance; when we are acquainted with their structure, and so familiar with their habits as to understand how it is they grow and live and multiply, and to what uses they may be applied, and by what contrivances, equally simple, invariable, and surprising, a small number of elements constitutes all those different organs, whose singular forms and brilliant colours so continually excite our admiration.

The power and wisdom of the Deity are proclaimed by no part of the Creation in more impressive language than by the humblest weed that we tread beneath our feet; but we must learn to understand the mysterious language in which we are addressed;

and we find its symbols in the curious structure, and the wondrous fitness of all the minute parts of which a plant consists, for the several uses they are destined for. This, and this only, is the "language of flowers;" and it was of this that I hoped in my former letters to give you some idea.

You tell me, however, that your curiosity is still unsatisfied, that you know not where to seek for other books in which similar information is to be found, and that the progress of your children in the classification of the various objects that surround them is accompanied by doubt and disappointment. I might easily name to you the very books you should read, and point out to you the very places in which you should search for the information you require, but I fear that you would still retain your opinion that it would have been better if, instead of idly referring you to the elucidations of others, I had had the industry to carry our correspondence a little further. Well then, let it be so! innocent knowledge is the best and most enduring foundation of permanent happiness, and far be it from me to refuse my humble assistance in contributing to the means by which the world may secure to itself the utmost amount of simple pleasures.

We will, therefore, resume our correspondence upon Botany; and this time, if your patience should not be exhausted, I promise to provide you with sufficient means for carrying your inquiries to whatever point you please, in respect of all those subjects which you would think of investigating for mere amusement's sake. In the first place, you shall have an account of those common tribes of plants about which I have, as yet, said nothing; then you shall learn to which of such tribes all the common plants both of the fields and the gardens belong; and finally, I will give you a little sketch of the general classification of those which have been previously brought before you only in detail. With this I may combine, when favourable opportunities occur, short episodes relating to the internal structure of plants, and the manner in which they grow; and, thus, I trust that an air of life and truth will be given to all the picture.

This will, I hope, fulfil your expectations, or, at least, relieve me from the reproach of unwillingness to satisfy your curiosity so far as my skill will permit me.

Suppose we begin with MAGNOLIAS, those beautiful American trees, which form the pride of European gardens, and the glory of the forests of North America, and many parts of Asia.

Observe that noble looking evergreen, with its large, shining, bright green leaves, in the bosom of which are reposing some cream-white flowers, much larger than any others you ever saw, and with a pile of purple and yellow stamens heaped up in their centre. That is the *Big Laurel* of the Americans, the *large-flowered Magnolia* (Magnolia grandiflora) of Botanists, and the handsomest of its tribe. It is found wild in the warmer parts of the United States, especially in South Carolina and the Floridas, and it shrinks from the cold weather of more northern climates. In its native forests it grows as much as ninety feet high, which is as high as the largest tree you ever saw in this country, and much higher than any even of the beautiful old elms that are scattered about in the park before you. A specimen of this size is described by a French Botanist as surpassing all other trees, "par son port majestueux, son superbe feuillage, et ses fleurs magnifiques." In this country it is too delicate to endure the blasts of our bitter winters, without some protection; but, as you see, it is very happy beneath the shelter of a wall, and pays no attention to the bonds with which it is secured to its prison. Beautiful as are its huge goblet-shaped blossoms, and surpassingly delicate as its buds of polished alabaster, it wants the rich perfume of many of its kindred. There is the *glaucous* Magnolia with smaller flowers, and leaves having a blueish bloom beneath them, by which nature points it out to the gatherers of the bark that cures the fevers so frequent in the unhealthy swamps where it delights to grow; and the *long-leaved Cucumber-tree* (Magnolia auriculata), so called because its leaves taste like Cucumber, with its spreading foliage, which has given it and some others the name of Umbrella trees; and the *long-leaved Umbrella tree* (Magnolia macrophylla), whose leaves are sometimes three feet long; these are species whose delicate cup-shaped flowers fill the air with their perfume. It is, however, in the East and not in the West that the Magnolia tribe has its fragrance most elaborated. In the *dwarf Talauma* of the Chinese (Magnolia pumila), with its yellow and brown flowers, and the *Tsjampaca*, the most beautiful of trees, beneath whose majestic foliage the native Indian constructs his cottage of Bamboo stakes and Palm leaves, the essence of the Magnolia perfume is developed in all its power. These trees are indeed the living altars from which a perpetual cloud of incense is ascending unto heaven day by day, as if in gratitude for the profusion with which the gifts of Nature are so prodigally poured forth from the lap of earth in those favoured regions.

After such an account as this you will be surprised to hear that Magnolias are nearly akin to the Crowfoot tribe (*Vol.* 1. *p.* 13. *t.* 1. 1.); that those beautiful trees, with their fragrant flowers and noble leaves, are related to such weeds as the wild Ranunculus, and the Thalictrum. And yet, such is undoubtedly the fact. Just observe the construction of the flower of this *heart-leaved* Magnolia (*Plate* XXVI. 1.). You see it has a calyx of three small reflexed sepals (*fig.* 1. *a.*); and six upright, yellowish, rather leathery petals, of which three are something narrower than the others. Within these are placed many stiff stamens (*fig.* 2.), arranged in several rows upon a receptacle of a somewhat conical figure (*fig.* 3. *a.*); each anther has two cells placed at the edge of a stiff fleshy filament (*fig.* 4.), and the cells are so situated that when they open, the pollen will fall out on the side next the petals (*fig.* 3.); this kind of anther is what is called technically *extrorsal.* In the centre of the flower is a large number of carpels, each of which contains one cell with two ovules in it (*fig.* 3.), and is terminated by a narrow thread-shaped stigma (*fig.* 3. *b.*). Those cells grow together into a solid pistil, and eventually change to a cone-like fruit, the seeds of which are principally composed of albumen, with a tiny embryo lying perdu in its base.

Such is the general structure of the heart-leaved Magnolia, and in what points of importance does it differ from a Ranunculaceous plant? It has a calyx of three sepals; so has Ranunculus Ficaria; it has six petals, so have many Anemones; its stamens are numerous, and placed on a receptacle beneath the carpels, their anthers grow to the edge of the filaments, and the carpels are very numerous; in all these things it agrees with Ranunculus itself; but the carpels grow to one another: the same thing happens in Love in a mist (Nigella); and, finally, the nature of the seed of a Ranunculus and a Magnolia is nearly the same. Are these plants then nothing but Crowfoots of a larger growth? merely Ranunculaceous plants with the stature of forest trees? Not quite so. The two orders are, as I have already stated, nearly akin to each other, but they belong to different races, and may be certainly enough distinguished. Do you see how each of these branches of the Magnolia is terminated by a little horn that springs from the base of the last leaf? (*fig.* 5. *a.*); that horn is a pair of stipules rolled together for the protection of the next leaf that is to be born; and that next leaf has a similar pair of stipules that roll up over the still younger leaf lying in its bosom; so that if you cut into the horn you will behold several generations of leaves lying enfolded the one within the other; this is the great mark of the Magnolia tribe, and enables you immediately to distinguish it, not only from the Crowfoots, but from most of those allied to it. And this is not only a curious but an important and highly interesting mark of distinction; the growing point of a branch of a Magnolia is tender, and requires to be carefully protected from the air, and from cold, and from those accidents to which all things must necessarily be subject that are directly exposed. To guard this tender part nature has many singular, but always most efficient, contrivances: in this instance the stipules are made to perform the business of protection.

The fruit of Magnolias differs in some respects from that of the Crowfoot tribe: especially in becoming large cones, from the back of which the seeds often hang down by long cords; but as Magnolias do not produce their fruit in this country it is unnecessary to describe this part of their structure.

Besides the plants called Magnolias, the curious *Tulip-tree* (Liriodendron tulipifera), one of the largest trees in the American forests, belongs to the present Order. You may know it by its singular truncated leaves, which look as if they were cut off at the end, and by its large pale green and purple flowers. It is not uncommon in the pleasure grounds of the old gentry of this country; some of the finest are to be seen at Sion, the seat of the Duke of Northumberland.

From these let us turn to a not less interesting, but more humble race of plants, of which the common

Barberry may be taken as the representative (*Plate* xxvi. 2.). This plant is so common in plantations and pleasure grounds, that all persons would be acquainted with it, if it were only for the quantities of bunches of red succulent acid fruit with which it is loaded in the autumn; and for its evil reputation as a poisoner of wheat when it grows in the hedge-row of a corn-field.

You will find the branches of this bush covered over with sharp spines (*fig.* 1. *a.*), some of which are divided into three, or five, or even a greater number of lobes, and some of which are undivided. What think you are these? Not prickles like those of the Rose, for they are regularly arranged over the stem, and will not break off by a slight pressure sideways; nor spines like those of the Hawthorn, for in the Hawthorn the spines originate in the bosom of leaves, but in the Barberry the leaves originate in the bosom of the spines. These parts are an exceedingly curious state of the leaf. They are the first kind of leaf that the Barberry produces when it shoots forth from the bud; but immediately after, or perhaps at the same moment with, their production, other perfectly formed leaves break out from their axils, and thus at nearly the same instant, the branches are covered with spines for their defence, and with leaves for their adornment. That these spines really are leaves you may easily ascertain by looking for a very vigorous shoot of the Barberry, when you will find some of them with the space between the stiff spiny lobes filled up by a web of parenchyma, others with the web hardly visible, and others with the spines alone remaining.

The leaves are themselves bordered by spiny teeth which are the points of their veins, and there is a little joint near their base (*fig.* 1. *b.*), by which they are articulated with their stalk.

From the midst of a cluster of leaves appear the yellow flowers, in a drooping raceme something like that of a currant. Each flower consists of three little external scales tipped with red; they are the outermost sepals; then of three petal-like parts (*fig.* 2. *a.*), the inner sepals; and within these of six genuine petals. The great similarity between the parts thus differently designated shews you that the distinction between a calyx and corolla is in many cases very arbitrary, although in other instances it may be plain enough. At the base of each of the true petals are two parallel yellow oblong glands (*fig.* 2. *b.*), the nature and use of which is unknown. Between these glands and opposite to the petals are the stamens, six in number, consisting of a filament somewhat thickened at its upper end (*fig.* 4. & 5.), and an anther whose lobes, growing to each side of the end of the filament, have a singular mode of opening. At first the lobes resemble those of any common anther, but when the time comes for the fertilization of the stigma, instead of splitting along the middle, the anther opens at the edge all round, except near the point, and liberates its valve or face, which curves back and allows the pollen to drop out (*fig.* 4. *a.*). This is a very curious phænomenon, and is technically called *bursting by recurved valves.*

The ovary is an oblong body (*fig.* 3.), terminated

by a flattish, round, sessile stigma, in the centre of which is a small opening that communicates with the single cell (*fig.* 5.) that the ovary contains. From the bottom of the cell, but rather obliquely, there arise two ovules (*fig.* 5.).

In time the ovary changes to an oblong acid scarlet fleshy berry (*fig.* 6.), containing one or two seeds (*fig.* 7.). The seeds have a tough skin, and enclose a slender embryo (*fig.* 8.), standing erect in the midst of hard albumen.

In this plant you will at once perceive several circumstances that you have not previously seen. In the first place its stamens are the same number as the petals, and opposite to them; and secondly their anthers open by recurved valves. These two points taken together, limit the Barberry Tribe, which contains the beautiful evergreen Ash-leaved species, or Mahonias, of which Berberis aquifolium or the Holly-leaved is so striking an instance, and also the singular brown-flowered Epimedium, whose small unattractive blossoms just raise themselves upon their thread-shaped stalks, and peep forth from the leaves which half shroud and half reveal them.

In the flower of the Barberry is a curious instance of irritability. The stamens are in a recumbent position when the flowers first open, lying back close-pressed upon the petals. But if you touch one of their filaments with a pin, the stamen gently rises up and strikes its anther against the stigma, just as the figures in old-fashioned clocks strike their hammers upon the bells when chimes are sounded. No one knows the cause of this curious habit; it is one of those certain but inscrutable facts, the explanation of which is probably beyond the faculties of man. There is one thing, however, connected with it that deserves to be noticed, although it does not throw light upon the nature of the phænomenon. If you dose the Barberry with laudanum or any opiate, the stamens are stupified and lose their elasticity; and if you poison the plant by some corrosive substance, such as arsenic, which produces inflammation in animals, a sort of vegetable inflammation is produced in the stamens of the Barberry. We are not, however, on that account to conclude that this plant approaches animals in its nature, but merely that the principle of life which pervades all nature is the same in its essence, and is affected in a similar manner by similar causes, whether it exists in an animal or a vegetable.

EXPLANATION OF PLATE XXVI.

I. The Magnolia Tribe.—1. A full-blown flower of *Magnolia cordata*, the natural size, showing the sepals *a*, and the six yellow petals, in the midst of which are seen the stamens and a small portion of the carpels.—2. The stamens without the petals, together with the mass of carpels in the middle.—3. A vertical section of the latter part, a little magnified; *a* shows the elevated receptacle, over the outside of which the numerous stamens are arranged; at *b b* are seen the stigmas with their uneven inner edge admirably adapted to collecting the pollen; and below some of the styles are the cavities of the ovary, in each of which are two ovules.—4. is a filament and anther, a little magnified.—5. The lower part of a leaf and its petiole, with its horn-like hairy stipule; at *a* is seen the scar of the opposite leaf which had dropped off, and *b* shews a portion of the end of the branch.

II. The Barberry Tribe.—1. A twig of the common Barberry (*Berberis vulgaris*), with a spine-like leaf at *a*; at *b* is a line showing where the leaf is jointed with its petiole; natural size.—2. A full-blown flower, magnified, showing the three inner sepals *a*, the six petals, each with a pair of parallel glands *b* at the base, the six stamens, and the central superior ovary; the outer sepals are too small to be seen in this direction, but they are visible upon the flowers in fig. 1.—3. A magnified view of the ovary, deprived of the floral envelopes, and shewing the origin of the six stamens.—4. The upper end of a stamen, magnified; at *a* is seen the singular mode in which the anther bursts by recurved valves.—5. A vertical section of the ovary, much magnified; the two erect ovules, and the open communication between the stigma and the cavity of the ovary, are plainly shewn.—6. A bunch of fruit, natural size.—7. One of the seeds, magnified.—8. A section of the same, shewing the dicotyledonous embryo standing erect in the midst of the albumen.

LETTER XXVII.

PISTIL—THE GOOSEBERRY TRIBE—THE VINE TRIBE.

(Plate XXVII.)

Is it possible that I should no where have told you the meaning of the common word *pistil?* You say you perfectly understand what a carpel is, but that you do not find in what respect a pistil differs from it. I am ashamed of my negligence, and hasten to repair it. The general name of the young fruit, consisting of ovary, style, and stigma, is pistil; the pistil is usually composed of several carpels, each of which has its own ovary, style, and stigma, as in a Ranunculus, where the mass of the carpels is the pistil; but it may consist of but one carpel, as in the Barberry, and in that case the words carpel and pistil have the same meaning.

Premising this, let me direct your attention to the GOOSEBERRY TRIBE, of which not only the plant that gives it its name, but all the currants are likewise members.

Currants you know are not confined to the kitchen garden; for besides the red, the white, and the black currants, every-body now possesses the sweet-scented yellow currant *(Ribes aureum)*, the crimson currant *(Ribes sanguineum)*, and other beautiful species which have been snatched from their native rocks and wilds

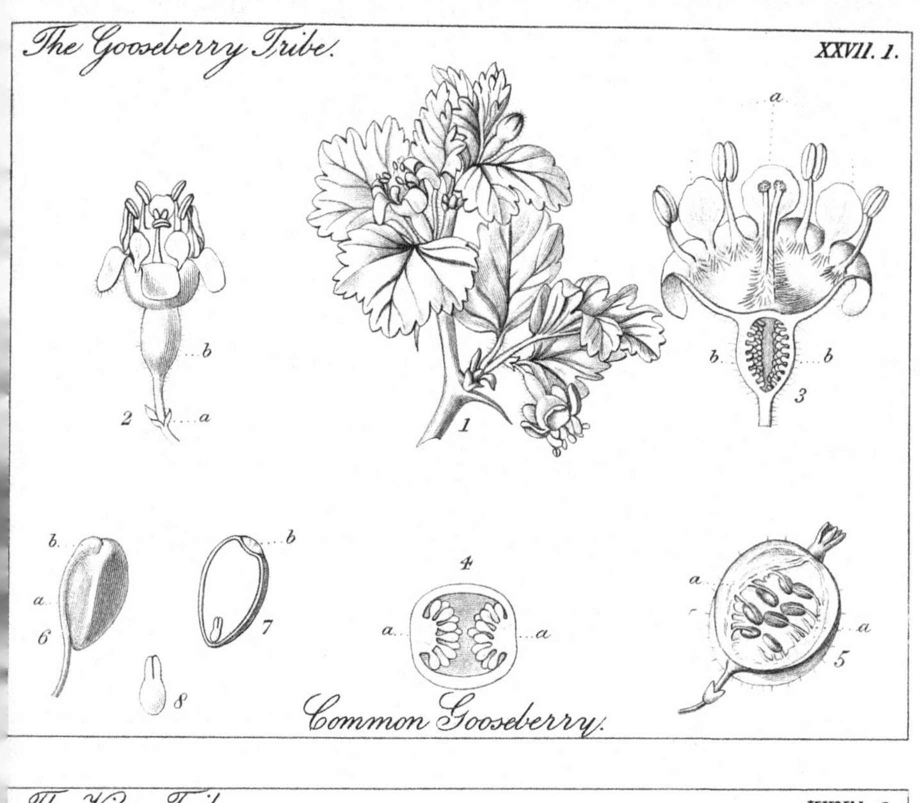
The Gooseberry Tribe.
XXVII. 1.
Common Gooseberry.

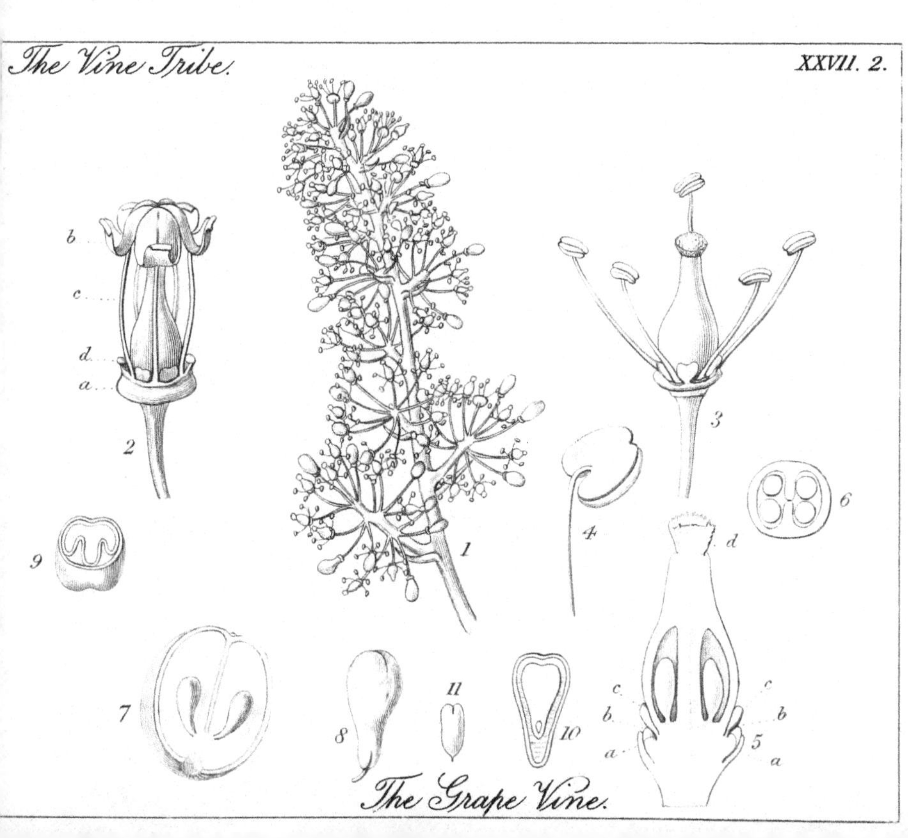
The Vine Tribe.
XXVII. 2.
The Grape Vine.

in New Albion to adorn the gardens of even English cottagers.

There is something in the organisation of these flowers particularly simple and pretty. Take the *Common Gooseberry*, for example (*Plate* XXVII. 1.). The spines with which the stem of this plant is defended, are of the same nature as those of the Barberry, that is to say they are rigid leaves, without the soft green pulpy substance, or parenchyma, that usually connects their veins. In the true leaves there is nothing to remark upon, further than that they are somewhat 3-lobed, and bluntly toothed along their margin; their stalks, however, are beautiful objects if examined by a microscope, because of the delicate border of half-transparent hair-like fringes, which, when magnified, look like the most brilliant needle-shaped crystals. The flowers are little green cups, purple in the inside, and grow in pairs, or singly, from among the leaves, which overshadow and protect them so completely that when a bush is in full flower you may pass it and hardly remark the blossoms. The cup (*fig.* 2.), green without and purple within, is the calyx; its border is divided into 5 blunt lobes, which are turned downwards. At the mouth of the cup you will find 5 tiny whitish scales, having each a short stalk with a tuft of hairs at its base (*fig.* 3. *a.*); these are petals. Between the petals are the stamens: 5 upright filaments, with an oval anther at the point, and a tuft of hairs at its base (*fig.* 3.). In the centre of all this apparatus rise two green threads, covered with long hairs at the base, but naked and terminated by a small stigma; of course these threads are styles: where then is the ovary? You shall see. Just below the cup of the calyx and above its stalk is a small oval swelling, clothed with long delicate hairs; it is hollow, and bears a great number of ovules, arranged in two lines upon its sides (*fig.* 3. *b.*); the styles are planted upon its summit; this then is the ovary, from which the gooseberry is to be produced. You could hardly have anticipated, before you began to study this science, how curious and complicated an apparatus is necessary for the production of so simple a fruit; everything you see is perfect, and in this tiny flower you have all the parts which you could find even in the gigantic Magnolia, only not so many of them, and differently arranged. And so it always is; be quite sure that in what may seem to you the most insignificant parts of the creation, there is the same foresight, the same admirable contrivance, the same beautiful adaptation of every part to the end it has to answer, and the same care to ensure against all accidents its multiplication after its kind, as in what we may habitually look upon so inconsiderately as the most perfect of the Creator's works. When rightly examined it will be found that no one thing is more perfect than another; each is perfect after its kind; imperfection is unknown in the creation; to argue otherwise is to argue against the power and wisdom of the Deity.

After a time the calyx-cup, the petals, stamens, and styles, shrink up and decay: at the same time the ovary swells, the hairs upon its surface either harden oı fall off, its interior becomes succulent, the ovules change to seed, they elevate themselves upon long stalks, and immerse themselves in the pulpy interior, the colour of the whole changes to red or yellow, and the ripe Gooseberry is completed.

If at this time it is divided into two parts from its apex to its base, it will be found to consist of a soft watery mass enclosed in a tough skin, which is the pericarp, and containing several hard seeds of a deep brown-purple colour, originating from a sort of web-like placenta. These seeds are secured by a green thread, which passes from one end to the other of the seed, on one side, forming a raphe (*fig.* 6. *a.*), and ending in a broadish expansion, or chalaza (*fig.* 6. *b.*). Within the skin, which is thick and tough, is a large quantity of hard albumen, at the base whereof lies a small dicotyledonous embryo (*fig.* 7.).

The common eatable Currants, and several other species found in different parts of Europe and Asia, are very like it; but this is not the case throughout the whole tribe. For example, the *Crimson Gooseberry* (Ribes speciosum), has a rich deep red calyx, with long narrow segments, and stamens projecting so far as to resemble those of the Fuchsia (*see Botanical Register, tab.* 1557.). In the *golden-flowered Currant* (Ribes aureum), the calyx is a bright clear yellow, with a long yellow tube, and the petals and stamens are short as in the common Gooseberry. These are the different forms of the Gooseberry tribe. Considering the manifest resemblance between a bunch of Currants and a bunch of Grapes, you will not be surprised at hearing that the Vine has some relationship with the Gooseberry tribe. This I now proceed to explain to you.

The common Vine, a native of the South of Asia, is the type of the VINE TRIBE *(Plate* XXVII. 2.). It has, as you know, very large lobed leaves, not at all unlike those of a Currant magnified, and its flowers grow in clusters, which however are not racemes, but panicles, that is to say, branched racemes (*fig.* 1.). The stem too is not that of a bush, but long and weak, and requires the support of other trees, to avail itself of which it is supplied with tendrils. Here let me pause to tell you what a tendril is; by its name you would suppose it some special kind of organ formed expressly for the purpose of helping the Vine to raise itself among the forests it naturally inhabits, and to ascend from the shady thickets where it is born, to the free light and air that are necessary for its existence. Not at all; this is not the plan of Nature. Plants are furnished with certain general parts, such as leaves, flowers, &c., and when any particular and unusual office is to be performed, some one of these parts is specially altered in order to meet the exigency. Thus in Combretum the stem is enabled to rise among other bushes by the soft and yielding stalks of its leaves being changed into stiff inflexible hooks; in the Sweet Pea the same office is performed by the principal leaf-stalk, which lengthens, branches, and twists itself round bushes and the branches of smaller shrubs. In some plants indeed this office is actually performed by the tips of the petals. In the Vine the arrangement is different from all those just mentioned, and equally simple; a considerable number of supernumerary panicles are prepared, on which no flowers are formed, but in their room a power of twisting round adjoining bodies is communicated to the branches; and these form what we call tendrils. But to return to other matters.

Each flower of a Vine (*fig.* 2.) consists of a calyx without any lobes to it (*a.*); five petals (*b.*) that hold together at the point, separate at the base, and are carried upwards with the extension of the stamens; of five stamens (*fig.* 2. *c.* and *fig.* 3.) opposite the petals, with long thread-shaped filaments and small oval anthers; of five glands alternating with the stamens (*fig.* 2. *d.*); and of a two-celled superior ovary, with a sessile roundish stigma (*fig.* 3.). In each cell of the ovary are two upright ovules (*fig.* 5. & 6.). The fruit is, as you know, a succulent berry, with one, or two, or three, or four, hard seeds nestling in the pulp (*fig.* 7.). These seeds are not a little curious; each has a pear-shaped figure (*fig.* 8.), and consists firstly of a tough external even coating, and secondly of a wavy bony lining, which does not follow the form of the outer skin, but puckers up, if I may so say, and forms a pear-shaped stone convex on one side, but with two deep furrows on the other, so that when you cut through it crosswise it looks almost like the letter T (*fig.* 9.). In the inside of the stone is a hard albumen, at the base of which (*fig.* 10.) lies a tiny embryo (*fig.* 11.).

This is the general character of the Vine tribe, the genera and species of which usually deviate so little from the Vine itself, that you would hardly fail to recognise them at the first glance. The *Fox grapes* of America (Vitis Labrusca and others) are, for instance, Vines with broader and more woolly leaves, and berries with a vile indescribable taste; the *River-grape* (Vitis odoratissima, or riparia), the delicious odour of whose flowers makes ample amends for their minuteness, would be taken for a common Vine if its leaves were not less lobed and more heart-shaped, and its berries so small, and black, and acid; while the *American creeper* (Ampelopsis quinquefolia), with its rich autumnal mantle of crimson, and the various kinds of Cissus, deviate from the ordinary appearance of the Vine chiefly in consequence of the leaves being separated into several distinct pieces.

Considering how common, and how useful a plant the Vine is, it is worth pausing here to consider a little, to what other plants it is most related; and I am the more inclined to do so because you like to be surprised, and some of its relations are undoubtedly of a very surprising character. What say you to Hemlock? I think I see you throw down my letter with what would be astonishment, if it were not for the incredulity mixed up with it. And yet I am not mystifying you, but in plain, sober, serious English, I say that the Vine and the Hemlock are nearly related to each other. For the proof;

The VINE *has*	*The* HEMLOCK *has*
Leaves deeply lobed, alternate upon the stem, with a stalk which is a good deal dilated at the base.	Leaves deeply lobed, alternate upon the stem, with a stalk which is a good deal dilated at the base.
A calyx with scarcely any lobes.	A calyx with scarcely any lobes.
A corolla with five petals.	A corolla with five petals.
Five stamens.	Five stamens.
A two-celled fruit.	A two-celled fruit.
Seeds with a very small embryo lying at that end of the albumen which is next the hilum.	Seeds with a very small embryo lying at that end of the albumen which is next the hilum.
An albumen deeply furrowed on the inside.	An albumen deeply furrowed on the inside.

In these points, which are of first-rate consequence, affecting the whole nature of the plants, you perceive that the two are the same. But

The VINE *has*	*The* HEMLOCK *has*
A superior ovary.	An inferior ovary.
Erect seeds.	Pendulous seeds.
Stamens opposite the petals.	Stamens alternate with the petals.
A pulpy fruit.	A dry fruit.

And some of these differences, slight as they are, are calculated to produce a considerable difference in the general aspect of the two plants, independently of the Vine being a woody climbing plant with panicled flowers, and the Hemlock a herbaceous biennial plant with umbelled flowers.

My proof of the relationship of the two plants does not however stop here, but is strengthened by other means. It is easy to shew a direct transition from the Vine to the Hemlock by a very brief examination of the plants that stand between the one and the other in a natural arrangement. Observe, the Vine is a Vitaceous plant, the Hemlock an Umbelliferous plant; to state this is to simplify the discussion.

Umbelliferous plants are allowed upon all hands to be distinguishable from Araliaceous plants, only by their fruit consisting of two parts instead of more, and by their fruit being dry instead of succulent.

Araliaceous plants are therefore Umbelliferous plants with succulent fruit. The common Ivy may be taken as a representative of the former. Many of the East Indian Ivies have their fruit in just such clusters as the Grape, and their leaves as much divided as in the Virginian creeper, so that they differ from Vitaceous plants only in their inferior fruit, pendulous seeds, and stamens alternating with the petals. Their close relationship is therefore unquestionable.

Then, if Vitaceous plants are closely akin to Araliaceous, and Araliaceous to Umbelliferous, it follows that Umbelliferous must be nearly allied to Vitaceous through Araliaceous, and consequently the Hemlock must be related to the Grape, as I at first told you. I hope you are now satisfied.

EXPLANATION OF PLATE XXVII.

I. The Gooseberry Tribe.—1. A twig of the common Gooseberry (*Ribes Grossularia*) in flower, natural size —2. A separate flower magnified, with the bractlets at *a*, and the inferior ovary at *b*.—3. The same divided in two, perpendicularly, and still more magnified; at the base is seen the one-celled ovary, with the wo parietal placentæ *b*.; *a* points to the petals.—4. Shews the appearance of a magnified transverse slice of the ovary, with the numerous ovules crowded over the placentæ *a a.*—5. is a section of a ripe fruit, shewing the remains of the flower adhering to its apex, and the seeds attached to the placentæ *a a*, by their long stalks.—6. is a magnified view of a seed with its stalk, raphe *a*, and chalaza *b*.—7. A section of the same, to shew the embryo lying in the base of the albumen; *b* is the chalaza.—8. is an embryo much magnified.

II. THE VINE TRIBE.—1. A portion of a bunch of flowers of the common Grape (*Vitis vinifera*).—2. A magnified flower in the act of opening; the calyx *a* is at the base in the form of a cup; the stamens *c* are pushing off the petals *b*; *d* are the glands of the disk.—3. Is the same flower after the petals have dropped off, and the stamens are liberated; here the ovary, with its sessile stigma, is distinctly seen in the middle.—4. A portion of a stamen magnified, with the anther in the position it occupies when heaving up the petals; see how it bends its shoulders (I beg pardon, its shoulder) to the task.—5. A magnified view of the longitudinal section of a young grape; *a* the calyx; *b* a ring from which the petals separate; *c* the glands; *d* the stigma. The ovules are standing erect in the two cavities.—6. A magnified view of a slice of the ovary, shewing that there are two ovules in each cell.—7. A section of a grape-berry.—8. One of the seeds.—9. A transverse section of the last, shewing the external coat and the internal stone.—10. A vertical section of the same, with the little embryo at the base of the albumen.—11. A highly magnified view of the dicotyledonous embryo.

LETTER XXVIII.

THE PITTOSPORUM TRIBE—THE MILKWORT TRIBE.

Plate XXVIII.

It is a common statement that New Holland produces no eatable fruits, for that even the few wild berries which the traveller meets with are more dry, tasteless and insipid than those of any other country. The Pears,* say the grumbling colonists, are made of wood, Cherries† have the stones on the outside of the flesh, Grapes‡ are nauseous, and grow on Bindweed, the Currant-bushes§ prickly, and the Gooseberries|| without thorns, while the Honeysuckle¶ has no odour, and the Oak** no foliage. Although these are mere idle tales, arising from the names of European plants being misapplied to New Holland species of a totally different nature, yet it is true that the whole of that vast continent is, as far as has yet been seen, destitute of any fruit-bearing plant that deserves cultivation.

The nearest ally of the Grape and the Currant for instance, is a beautiful twining evergreen plant with small dark green leaves, and large berries of the deepest

* Xylomelum pyriforme. † Exocarpus cupressiformis.
‡ Polygonum adpressum. § Leucopogon. || Gaultheria.
¶ Banksia. ** Casuarina.

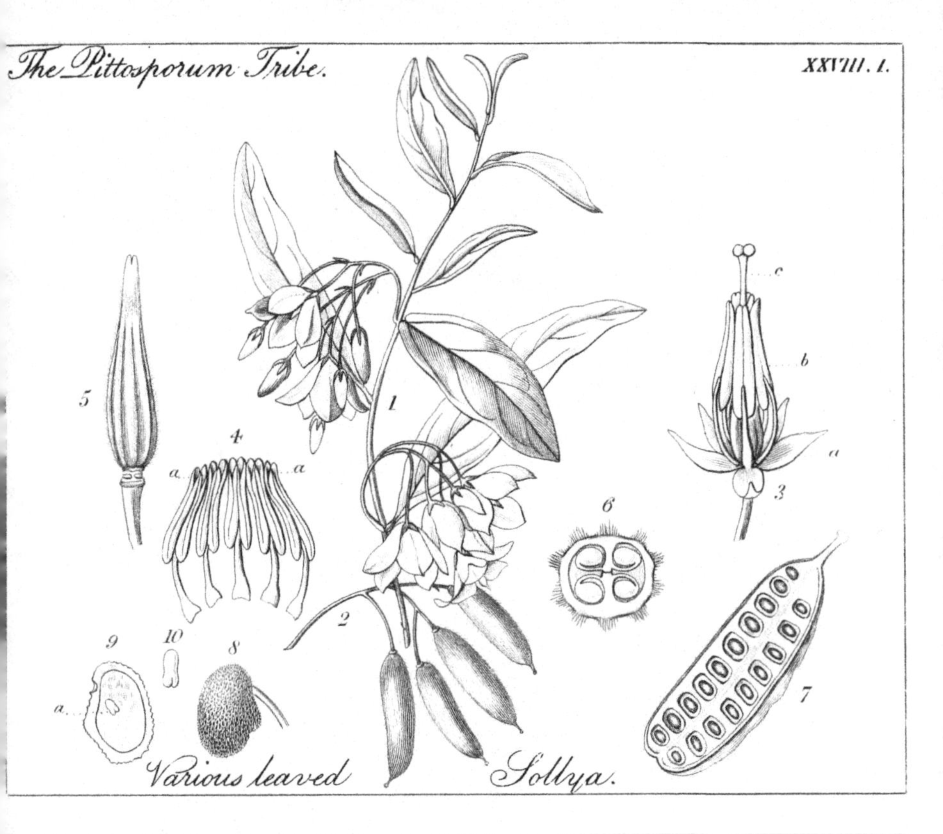
The Pittosporum Tribe.
XXVIII. 1.
Various leaved Sollya.

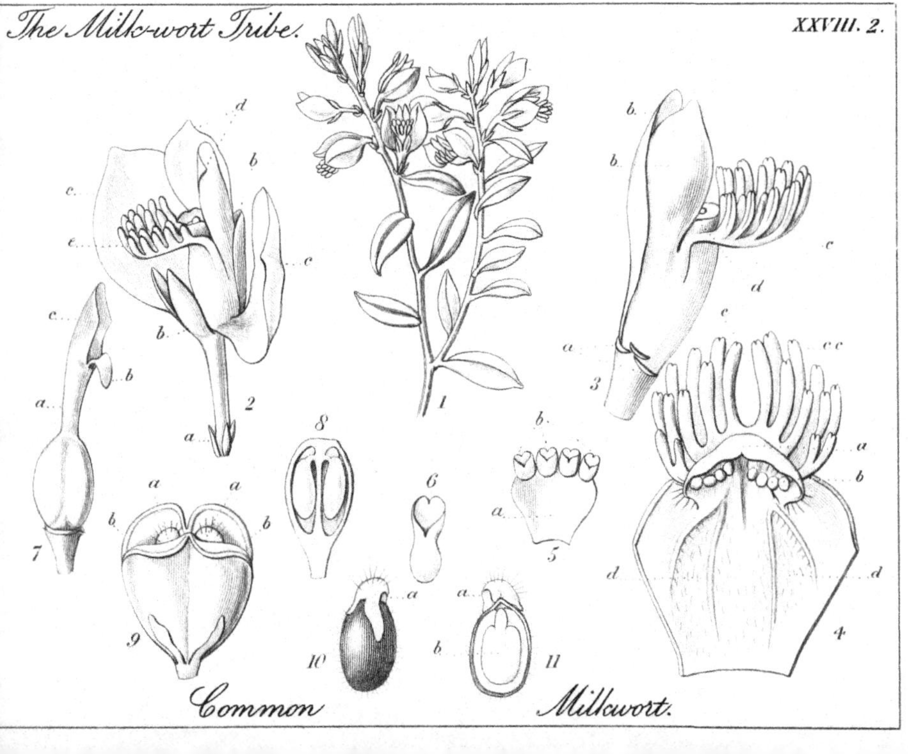
The Milk-wort Tribe.
XXVIII. 2.
Common Milkwort.

lapis lazuli blue, resembling nothing so much in colour, and, to appearance, even in texture, as the fine deep blue of the porcelain of Sévres. It has small greenish-yellow bell-shaped flowers, and Botanists call it Billardiera; in Van Diemen's Land, where it is extremely common, the name of *Apple-berry* is applied indiscriminately to all the species. This most lovely fruit, although, as I shall presently shew you, nearly akin to the Vine, has none of its delicious flavour, but is a mere mass of cottony, or rather spongy, dry pulp, which would be tasteless if it were not for a dash of turpentine which is perceivable. If you do not possess the Billardieras already, let me advise you immediately to procure them for the low treillage in front of the conservatory, where they may be kept very well in mild winters, if protected with a little straw or a mat.

Sure I am that if you do not possess these, you are, at least, the mistress of a plant of Sollya, and this will do as well as a Billardiera for the purpose of studying the characters of the PITTOSPORUM tribe to which they both belong.

Sollya heterophylla (Plate XXVIII. 1.) is a little evergreen climbing plant, with light-green, alternate, oval, shining leaves, most of which have an even edge, but a few are, now and then, serrated; there are no stipules, and the leaves have a slight odour of turpentine when rubbed. The flowers grow in small nodding cymes or clusters, are bell-shaped, and of a beautiful bright blue, not unlike that of our own wild *blue-bells* (Hyacinthus non scriptus). The flower-stalks are slender, and have each a tiny awl-shaped bract at the base. The calyx consists of five narrow sharp sepals (*fig.* 3. *a.*), within which arise the ovate petals. There are five stamens alternate with the petals, and growing from below the ovary (*fig.* 3. *b.*), with narrow yellow anthers, converging into a cone that surrounds the style, and opening by two pores at the point (*fig.* 4.). The ovary is a slender downy body (*fig.* 5.), furrowed with deep channels, narrowed into a smooth white style (*fig.* 3. *c.*), ending in a small two-lobed stigma, and containing two cells (*fig.* 6.), in each of which are two rows of ovules attached to the placenta by long slender stalks; these ovules are separated from each other by thin green horizontal plates. The beauty of Sollya lies entirely in its flower; its fruit is not rich and tempting in appearance like that of Billardiera, but an oblong, hard, hairy, brownish body (*fig.* 2 & 7.), tipped by the hardened style. If you cut across it you will find it apparently separated internally into four cells; but this is a deception, arising from the matter that lines the inside of the ovary having grown up between the ovules so as to lodge them all in distinct cells; this is most evident when the fruit is cut lengthwise (*fig.* 7.), when every one of the little brown seeds is seen nestling in its own close box. The seeds have a minutely rough skin (*fig.* 8.), and retain the stalk you saw on them when ovules (*fig.* 8. *a.*). If you open them skilfully they will be found to consist of some hard brownish albumen, in which a very small embryo (*fig.* 9. *a.*) is lodged near the hilum.

At first sight a plant like this appears very unlike a Vine; but if the two are botanically contrasted, it will be obvious enough that they are in fact very nearly related. Sollya and Billardiera climb; so do Vines; they have all alternate leaves without stipules, their stamens are 5, their petals 5, their ovary superior and two-celled, their embryo a minute body lying in albumen; and the Vine and Billardiera agree in having soft fruit, not that that is of much importance. These points of resemblance are so numerous, among the most important parts of the structure, as to render the relationship of the tribe before us and Vines unquestionable. They differ, however, too much to be actually included in the same tribe; for these plants have not stamens opposite the petals, nor erect seeds, nor glands below the ovary, all which are distinctive marks of the Vine tribe. They have therefore been collected into an assemblage called the PITTOSPORUM TRIBE, after a genus of which no mention has as yet been made, and which you do not often meet with in gardens. Its species are very different in habit from Sollya and Billardiera, being upright evergreen bushes, and not climbers, and having a capsule that opens into valves, and not a soft berry. The most common of the genus is the *Tobira Tree* (Pittosporum Tobira), an evergreen laurel-like bush, with cream-coloured sweet-scented flowers. It is not rare in extensive collections, and in some warm situations will even grow in the open air without protection in the winter. Nothing can be more unlike a Vine than Pittosporum itself; but it is closely allied to Sollya, which is next akin to Billardiera, the affinity of which to the Vine has been demonstrated.

The last case has served to shew you another instance, in addition to those you are already acquainted with, of plants, apparently very dissimilar, being in reality near relations, and that it is only to Botanists that the links which hold together what is, not very correctly, called the mighty chain of the creation, are perceptible. It will not be uninteresting to take this opportunity of making you acquainted with a highly curious natural order, which, with far more apparent resemblance to the Pittosporum tribe than the Vine, has in reality a much more distant relationship.

On heaths, and sunny knolls, and on many a naked down all over England, is found a pretty little herb, with exquisitely curious tiny blossoms of blue, or white, or pink, which modestly peep up from the turf that cherishes them. They call it MILKWORT (*Plate* XXVIII. 2.). The ancients fancied that it, or some such plant, possessed the property of increasing the quantity of milk in the cows that fed upon it; hence its name. One never sees it cultivated in gardens, and yet it is of an exceedingly beautiful, and most curious structure; but its flowers are so small that all which is most admirable in it is overlooked by the incurious observer, and larger foreign species, chiefly from the Cape of Good Hope, are nursed in greenhouses in its room. Our Milkwort (Polygala vulgaris), has weak rambling stems from two to eight inches long, clothed with minute, oval, sharp-pointed, deep green leaves, and terminated by a short raceme of flowers. These have so very uncommon a form that I must describe them more particularly than usual.

Separate a single flower from the others (*fig.* 2.). At the base of its stalk grow three little scale-like bracts, of a pale delicate lilac colour, like the stalk itself (*fig.* 2. *a.*). The calyx has five sepals—"Five?" you will say, "I see but three."—The calyx *has* five sepals, of which three are small, green, and narrow (*fig.* 2. *b. b.*), and two broad, bright blue, and spreading away from the flower like wings (*fig.* 2. *c. c.*). You no doubt mistook the last for petals, because they were delicate in texture and rich in colour; but it is not such qualities that constitute a calyx, as you have long since been aware; a calyx is merely the outer row or whorl of leaves, and you will find that the two blue wings of the Milkwort grow from between the green sepals out of a row (*fig.* 2.), which, although a little broken, evidently belongs to one whorl.

Now it is a general rule that whatever number of sepals there may be in a flower, there will be the same number of petals, if there are any at all; and although no doubt we have exceptions to this, yet such is the rule in most cases. The Milkwort looks as if it were one of the exceptions, for upon examining its corolla, the greatest number of parts you seem able to make out is three (*fig.* 3. *b. b. c.*). Strip off the sepals, noting carefully the spots from which they separate (*fig.* 3. *a.*); you will then have a corolla with two erect, lanceolate, blue segments (*fig.* 3. *b. b.*), and a sort of fringed projection in front of them (*fig.* 3. *c.*), called the crest; this is but three parts. Let us, however, examine the beautiful little crest a little more particularly, for which purpose we will cut it off the back petals, and look at its inside (*fig.* 4.); we shall then find that it is a light blue, downy plate, divided at the point into two parcels of fringe (*fig.* 4. *c.* & *c. c.*), within which there is a little hood (*fig.* 4. *b. b.*), having the most delicate little whiskers in the world at its base. What are the two bundles of fringe, and the little hood? they must be something similar in nature to what is found in other flowers, although strangely disguised. Botanists say that the hood is the point of the middle petal of three, and that it has the two side petals with their fringes firmly attached to its back, so that the crest is in reality made up of three petals naturally soldered together, and these, together with the two other petals at the back, make up the number five of which we have been in search.

But where are the stamens of this curious plant? Not at the base of the ovary (*fig.* 7.), nor attached to the calyx, nor any where within sight. Lift up or press back the hood we have been talking of, and there you will find them. There are two rows of little yellow cases hidden beneath this hood (*fig.* 4. *b.*), four cases in each row, and adhering to a thin membranous plate (*fig.* 5. *a.*); the latter is the united filaments, and the cases are the anthers. Why they thus lie perdu beneath the hood in the inside of the crest you will perceive presently; in the mean while observe that each anther not only opens by a pore at the point, (*fig.* 6.) but is one-celled. The ovary (*fig.* 7.) is an oblong body, containing two cells, in each of which is one pendulous ovule (*fig.* 8.); it is furnished with a club-shaped style, and a thick two-lipped stigma, the upper lip of which (*fig.* 7. *c.*) is purple, large, and hooded, the lower (*fig.* 7. *b.*) small, flat, yellow, and bent downwards. All the parts of the flower are so placed about the stigma, pressing upon it, that there is no room for insects, or even wind, to insinuate themselves for the purpose of dispersing the pollen; on that account the stigma fronts the hood under which the anthers are hidden, and, opening its wide mouth, (for surely that may be called wide, the two lips of which are so far apart as in this plant (*c.* & *b. in fig.* 7.),) gapes to receive the pollen, which easily falls into it when the anthers open. The fruit is a heart-shaped capsule (*fig.* 9.), opening through the middle of the cells, and allowing two pendulous seeds to fall out. The latter (*fig.* 10.) are small, oblong, dark brown, hairy bodies, at the hilum of which there is a curious white hairy lobe, or *caruncula* (*fig.* 10. 11. *a.*). They contain a large, flat, dicotyledonous embryo, lying in a small quantity of albumen (*fig.* 11.).

The Milkwort Tribe obviously differs in so many respects from the Pittosporum Tribe that it would be tedious and unnecessary to recount them. Neither is there any other assemblage of plants sufficiently similar to be mistaken for them, unless it is the Pea Tribe (Letter VIII.), and with that students do sometimes confound them, because of the resemblance that the flowers of the Milkwort appear to bear to what are called papilionaceous. If, however, they are attentively considered they will be found not to resemble them in reality, for the two wings, which might be mistaken for the wings of a papilionaceous flower, belong to the calyx and not to the corolla, which is a most important difference.

Many a plant belonging to the Milkwort tribe grows wild in the southern parts of Europe, and at the Cape of Good Hope; nor are species altogether wanting in any quarter of the globe. The Cape kinds are, as I have already told you, often cultivated in Greenhouses, of which they are a great ornament. Generally these plants are bitter; but some of them abound to such a degree in saponaceous properties as to be real vegetable *blanchisseuses*. There is, in particular, a plant in Peru, called *Yallhoy* (Monnina polystachya), an infusion of whose bark is used by the ladies of that country for washing their beautiful hair, and finer is that hair said to be than any other in the world. This I am not so unjust as to believe; but the mere statement, with all its exaggeration, suffices to shew that the plant in question possesses properties of no common kind.

EXPLANATION OF PLATE XXVIII.

I. The Pittosporum Tribe.—1. A twig of *Sollya heterophylla*, or *the various-leaved Sollya*, in flower.—2. A small cluster of its fruit.—3. A calyx magnified, with the stamens converging in a cone around the style; *a* the sepals, *b* the anthers, *c* the style.—4. A set of the stamens curved back, and opened out; *a* the pores by which the anthers discharge their pollen.—5. An ovary.—6. The ovary cut across transversely, exhibiting the ovules lying in the two cells, and the ten ridges of hair that clothe the surface of the ovary.—7. A longitudinal section of a ripe fruit, shewing how the seeds are lodged in separate hollows, produced by the growing up of the sides of the ovary.—8. A seed, with its stalk or funiculus, *a*.—9. A section of the same, with the embryo, *a*, lying in discoloured albumen.—10. An embryo very much magnified.

II. THE MILKWORT TRIBE.—1. A twig of *common Milkwort* (Polygala vulgaris).—2. A complete flower much magnified; *a* the bracts, *b b* the small sepals, *c c* the petaloid sepals, *d d* the back petals, *e* the crest.—3. A corolla from which the sepals have been removed; *a* the scars from which the sepals have been taken, *b b* the back petals, *c* the crest, *d* the hood lying within the pouch.—4. A portion of the crest very much magnified, and seen from the inside; *a* the hood or middle petal, *b b* the stamens, *c* one set of fringes, or one of the side petals, *c c* the other set of fringes, or side petal, *d d* the hairs on the inside of the crest.—5. One of the two parcels of stamens, *a* the filament, *b* the anthers.—6. An anther.—7. An ovary, with the style *a*, and the stigma *b c*.—8. A longitudinal section of the ovary, shewing the two pendulous ovules.—9. A ripe capsule, opening and exposing its seeds *a a*, between the valves *b b*.—10. A seed; *a*, its caruncula.—11. The same cut lengthwise, shewing the embryo *b*, and the caruncula *a*.

LETTER XXIX.

THE MIGNONETTE TRIBE—DISK—THE CAPER TRIBE.

Plate XXIX.

ONE of the first flowers that we learn to gather—the very last that we cease to value—is Mignonette, that simple, unattractive weed, which is the envy of the gay and glittering throng that surrounds it in a garden, and which has no rivalry to dread, except from the Rose and the Violet. We are delighted with its fragrance, but we seldom think of asking whether, beneath the green and brown colours of its flowers, there may not lurk some hidden beauties equally deserving of admiration. It is one of the advantages of Botany, that it of necessity leads us to such inquiries. Let us look into its history and structure.

Mignonette (Reseda odorata) is generally reputed to be a native of Egypt and Barbary; but the only certain station for it is in the sandy country about Mascara, a fortified town of Algiers; writers on the Botany of Egypt make no mention of it. We find it in our gardens to be annual, sowing its seeds spontaneously, and springing up year after year wherever it has once been cultivated; but in reality it is a half-shrubby plant, like a wall-flower, and will live a long while, if protected from cold in the winter. I once

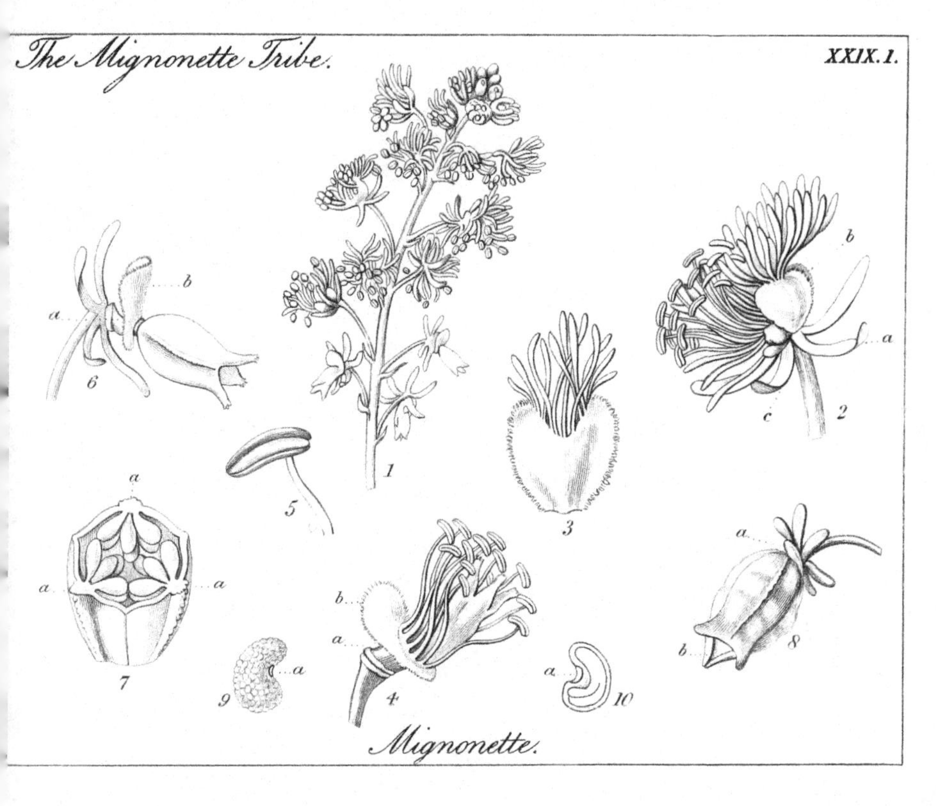
The Mignonette Tribe.
XXIX. 1.
Mignonette.

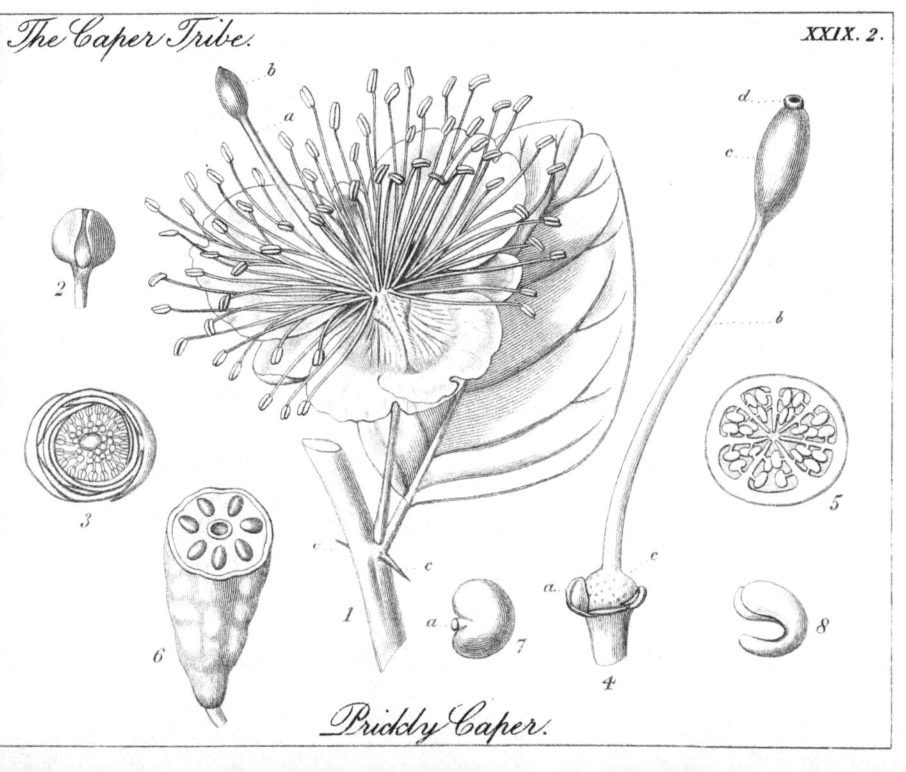
The Caper Tribe.
XXIX. 2.
Prickly Caper.

knew of a plant which had established itself in a crevice at the top of the back-wall in the inside a greenhouse, just beneath the glass roof; it remained growing in that situation for some years, putting forth its odoriferous flowers the whole winter long; and it had become quite a bush at the time when it was destroyed by an accident.

In the leaves of Mignonette there is nothing sufficiently remarkable to point out; but the flowers are exceedingly curious. They grow in racemes (*Plate* XXIX. 1. *fig.* 1.), on longish stalks, from the bosom of little green bracts. Each consists externally of a calyx, composed of six, linear, green sepals (*fig.* 2. *a. a.*), of equal size, and rather shorter than the petals. The latter are also six in number, but very unequal in size and dissimilar in form; the largest (*fig.* 2. *b.* and *fig.* 3.) are green, fleshy, wedge-shaped bodies, bordered with unequal, whitish, gland-like hairs, and having at the upper end a crest, consisting of white, flat threads, which are broader at the upper than the lower end. The smallest petals are roundish, and much shorter than their crest-like appendage, which, moreover, is made up of much fewer parts than that of the largest petals. From within the base of the petals there rises a short green stalk (*fig.* 6. *a.* and *fig.* 4. *a.*), called the gynophore, from the top of which springs a one-sided, brown, hairy lobe, or *disk* (*fig.* 4. *b.* and *fig.* 6. *b.*), hollowed out into a short tube at the bottom, where it surrounds the base of the ovary, and bearing twelve stamens at the top of the tube (*fig.* 4.).

Before we proceed further, let me detain you a moment with the meaning of the word DISK: a term that has just occurred. We formerly had it as the name of the central part of a compound or rather a composite flower. (Vol. I. p. 203.), comprehending all the florets which have a tubular structure with an equally divided border; in the present instance it is used in a different sense. It means a supernumerary organ, different from the stamens or petals, and originating at the base of one or other of them. Nothing can well be more variable in its nature than this disk; in the Mignonette it is, as you see, a one-sided, hairy lobe; in some plants it is a fleshy ring surrounding the ovary; in others a small number of glands in the same place; in Black Horehound you formerly saw it in the state of a green fleshy base to the lobed ovary (Vol. I. *Plate* XVI. 1. *fig.* 4. *a.* and 6. *a.*); and in the poppy-flowered Pæony you will find it constituting a deep purple case, enveloping the ovaries, and cut into irregular segments at its edge. In all these instances the disk is considered to be in reality either corolla, or stamens, in a disguised state; in the example before us, it is to be referred to the corolla.

The ovary of the Mignonette (*fig.* 6.) is an oblong, three-cornered, three-horned, one-celled case, having its horns terminated by the stigmas, and its ovules arranged in triple rows upon three narrow placentæ (*fig.* 7. *a. a.*), corresponding with the principal angles of the ovary. If viewed with a magnifying glass, the angles will be found covered with a cold-grey frost, of an extremely pretty appearance.

The seed-vessel of the Mignonette is an oblong brown case (*fig.* 8.), opening at the point into a triangular passage, through which the seeds readily fall out. The seeds (*fig.* 9.) are brown, warted, kidney-shaped bodies, attached by the middle of their concave side (*fig.* 9. *a.*), and contain an embryo, which is curved like the seed itself (*fig.* 10.).

Besides Mignonette, the genus Reseda contains many other species; they are all, however, confined, when cultivated, to Botanic Gardens; for they are but little superior in external appearance to the Mignonette itself, and they have none of its fragrance. Two of the species are wild in Great Britain, and one of them (Reseda luteola), the *dyer's weld*, possesses the property of imparting a beautiful yellow colour to linen and wool.

You have remarked, that in Mignonette the ovules grow to the shell of the ovary, and not to the middle; a similar circumstance has been pointed out to you in the tribes of the Violet, the Poppy, the Passion-flower, and others, formerly brought under your notice. I think I have somewhere already told you that the place where the ovules adhere to the ovary is called the *placenta;* and that the manner in which they adhere is hence called their *placentation;* let me now add, that when the placentæ are upon the shell of the ovary, as in this and the other instances already alluded to, the placentation is technically called *parietal;* I mention this, because the latter term is of such common occurrence that Botanists are obliged to have recourse to it frequently. You will understand this readily enough if you compare with each other *Plates* I. 2. *fig.* 6.; IV. 2. *fig.* 6.; and V. 1. *fig.* 4.

Should you now seek to discover some tribe of plants with which the Mignonette can be identified, you would undoubtedly fail, for it is extremely unlike any of those hitherto mentioned to you by me. On this account it forms a group by itself, called Resedaceæ, or the MIGNONETTE TRIBE. There are, however, plants allied to it by many important characters, the most interesting being what are popularly called Capers. We will now investigate their structure.

The CAPER TRIBE (Capparidaceæ), may be considered as represented by that species which furnishes the Capers sold by the Italian oil-men. This plant (Capparis spinosa, *Plate* XXIX. 2.) inhabits the chalk and volcanic rocks of the South of Italy and Sicily, especially those within the influence of the sea; there it enjoys a bright warm summer and a mild and equable winter, and trailing over the precipices that it inhabits, gives to the wild and rugged scenery a summer charm which the Myrtle and the Rock-rose in vain attempt to emulate. Wherever a similar climate can be found, the Caper bush is transferred for cultivation, on account of the mild, agreeably pungent properties of its flower-buds. It is these which form the Capers of the shops, their quality depending upon the age at which they have been collected; the youngest, and consequently the smallest, forming samples of the best, and the largest and oldest of the worst quality. But let us examine the Caper plant more systematically.

It is an undershrub, with long, smooth, shining, trailing, purple branches, bearing alternate, ovate, flat, dull green leaves, edged with purple, and placed upon a short purple stalk. At the base of the stalk, on each side, is a short straight spine, supposed to be a disguised stipule. From the axils of the leaves the flowers (*fig.* 1.) grow singly, on hard, smooth, purple stalks. They have four, spreading, oblong, obtuse, concave sepals ; four white petals, notched at the end, downy at their base, and so placed that two adhere to each other, as if really united ; there is a large number of stamens growing from the base of a central column, with thread-shaped filaments ; and, finally, the ovary (*fig.* 1. *b.*) is an oval purple case, growing on the end of a long cylindrical gynophore (*fig.* 1. *a.*). The interior of the ovary (*fig.* 5.) is very like that of the poppy (*Plate* I. 2. *fig.* 6.), having several plates covered with ovules, projecting from the shell, and not meeting in the middle : the placentation being therefore parietal. The stigma is a roundish, sessile, purple tip to the ovary (*fig.* 4. *d.*). At the base of the gynophore, on one side of a sort of cushion that bears the stamens, is a small, ovate, convex, gland-like disk (*fig.* 4. *a.*). When the fruit is ripe it becomes an oblong, knobby body (*fig.* 6.), filled with firm pulp, within which the seeds lie in as many rows as there previously were placentæ. The seeds themselves are very like those of Mignonette, only smooth, not warted.

I have already said that the Capers of the shops are

the unexpanded flower-buds of this plant (*fig.* 2.). If you cut them across you will find their appearance in a transverse section sufficiently curious. They consist of several green leaves wrapped one over the other (*fig.* 3.), and enclosed within a couple of concave bracts; within these lie the petals, enwrapping the stamens, which are closely packed round either the gynophore or the ovary.

Our gardens contain nothing included in the same group as the Caper, except certain annuals called Cleomes, a few of which have gay starry flowers, and long stamens, far less numerous than in the Caper itself.

It is obvious that this plant accords with the Mignonette tribe more than any others yet examined. It has, independently of its polypetalous flowers, a considerable number of stamens, a disk adhering to the part in which the stamens originate, a gynophore on which the ovary is elevated, an ovary with parietal placentation, and kidney-shaped seeds, with a curved dicotyledonous embryo. These circumstances undoubtedly indicate a near alliance between the Caper and Mignonette, and, in reality, the general opinion now seems to be in favour of their standing next each other, only in distinct groups.

With regard to *Cleomes,* I must refer you to the Hothouse for information concerning them. They are considered to stand, as it were, between the Caper tribe and the Cruciferous tribe (Vol. I. *p.* 55.); connecting, in a very conspicuous manner, plants that otherwise would not have been readily brought near each other.

EXPLANATION OF PLATE XXIX.

I. THE MIGNONETTE TRIBE.—1. A few flowers of Mignonette (*Reseda odorata*).—2. A perfect flower magnified; *a a* sepals; *b* the upper and larger petals, with their crested appendages.—3. One of the upper petals still more magnified.—4. A flower with its sepals and petals cut off, shewing at *a* the gynophore, and at *b* the disk, with the stamens and ovary within them —5. The upper end of a filament, with its anther.—6. A view of the young ovary, when the petals and stamens have dropped off; *a* the gynophore, *b* the disk.—7. A transverse section of the ovary, with the ovules adhering in triple rows to the three parietal placentæ.—8. A ripe fruit, opening by a triangular passage, *b*, at the apex, and having the remains of the disk, *a*, adhering to its base.—9. A ripe seed; *a* the scar.—10. A longitudinal section of the same, with the dicotyledonous embryo; *a* the scar.

II. THE CAPER TRIBE. — 1. A twig of the prickly Caper (*Capparis spinosa*) in flower; *a* the gynophore, *b* the ovary, *c c* the spiny stipules.—2. A young flower-bud, in the state in which it is gathered for pickling.—3. A transverse section of the same, magnified. —4. A view of *a* the disk, *b* the gynophore, *c* the ovary, *d* the stigma, *e* the receptacle of the stamens magnified.—5. A transverse section of the ovary, with the ovules adhering to the plate-like parietal placentæ. —6. A portion of a ripe fruit cut across.—7. A ripe seed; *a* the scar. —8. An embryo extracted from the seed.

LETTER XXX.

THE CACTUS TRIBE—THE GOURD TRIBE.

Plate XXX.

Besides the plants spoken of in my last letter, there are several others whose placentation is also parietal (see page 39.), and it will be better, before we proceed to other subjects, to examine some of them; especially two which are of very common occurrence.

The plants called Cactuses, which, from the profusion of large richly-coloured flowers that some species are loaded with, have given to our conservatories an air of magnificence which was quite unknown till of late years, constitute the small group of Cactaceæ. The species are in all cases succulent, and with the single exception of the Pereskias, destitute of leaves, in whose room the stem is either green and leaf-like, or at least covered over with a green integument, which has the structure of the pulpy part of a leaf, and like it executes the office of respiration. You will form a general idea of this highly curious natural order when you are told that the plants called *Indian Figs* (Opuntia), with their prickly, jointed, flattened stems, on which the Cochineal insect feeds; *Torch-thistles* (various species of Cereus), whose angular trunks rise erect and singly into the air, like fantastic vegetable

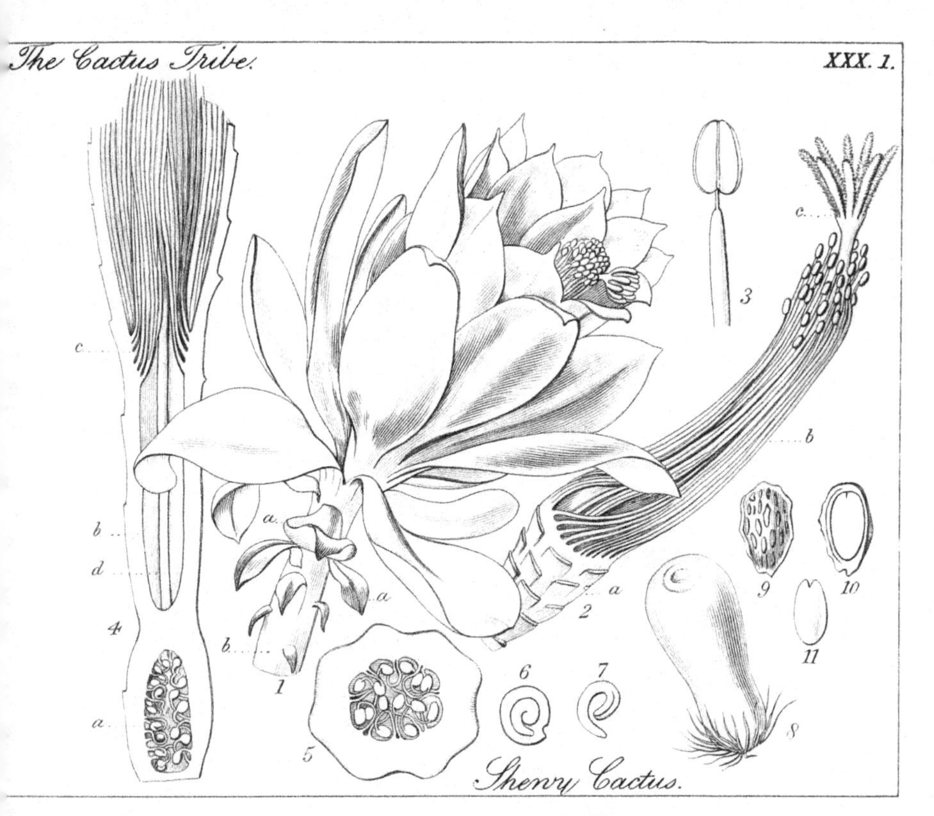
The Cactus Tribe.
XXX. 1.
Showy Cactus.

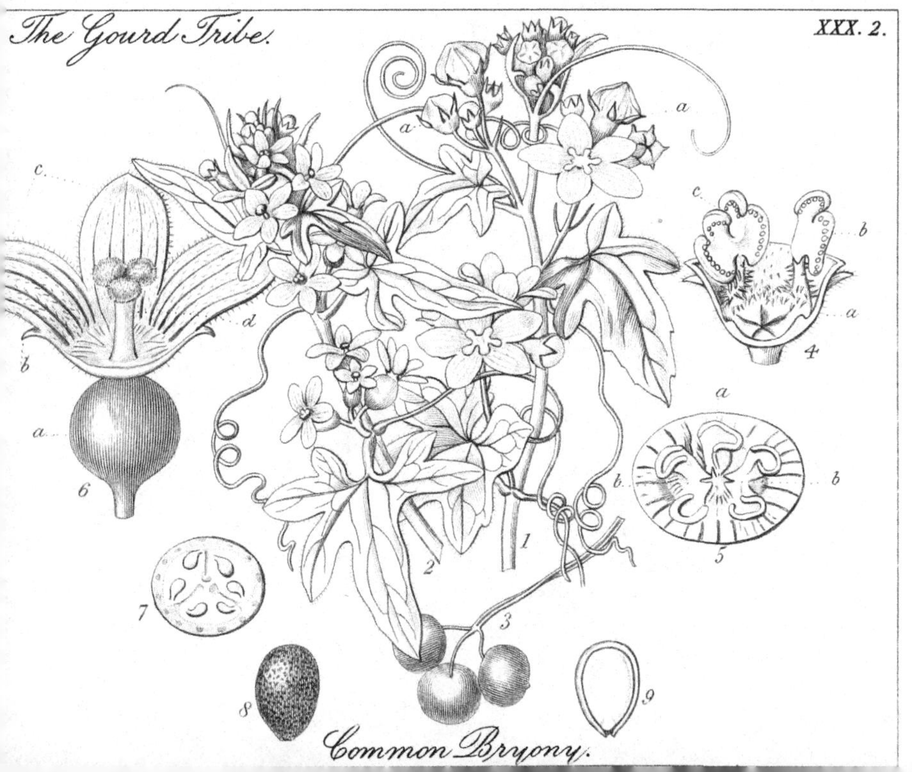
The Gourd Tribe.
XXX. 2.
Common Bryony.

columns; *creeping Cereuses*, with their long pendent branches, which might be taken for the tails of some animal, if it were not for the gay, rose-coloured flowers they push out from time to time; and all the strange races of *Melon-thistles* (Melocacti), *Porcupine-thistles* (Echinocacti), and *Hedge-hog thistles* (Mammillarias), whose names sufficiently attest their extraordinary appearance—I say, you will form a clear general idea of this curious Cactus tribe, when you have collected in your mind all the remarkable plants that have now been named; and I cannot anticipate any difficulty in your doing so, because they are among the commonest plants that inhabit greenhouses. All these species are destitute of true leaves, except when they are first beginning to grow. Just at that time they do indeed produce little succulent bodies, which we know to be rudiments of leaves; but such parts drop off soon after they are born, and the only representatives they leave behind are the stiff, hooked spines, with which so many species are covered. The parts which are mistaken for leaves in the Indian fig, or some of the more common species of Cereus, are only the flattened joints of the stem.

It would be difficult to find any race of plants, where a more obvious connection exists between the manner in which they are constructed and the situations it is their destiny to live in. The greater number grow in hot, dry, rocky places, where they are exposed for many months in the year to the fiercest beams of a tropical sun, without a possibility of obtaining from the parched and hardened soil, more than the most scanty supply of necessary food. Under such circumstances plants of an ordinary structure would perish; but Cactuses have a special power of resisting heat and drought, and, like the Camel, they carry with them a supply of water for many, not days but, months. It usually happens that once a year, during several weeks at least, the air that surrounds them is saturated with moisture, and the soil they live in is drenched by ceaseless rains. At this time they grow fast, all the little cavities in their tissue, of which there are countless millions, are filled with liquid nourishment, and they may be literally said to gorge themselves with food. Then, when the rains cease, and the air dries up, and the Spirit of the desert reassumes his withering dominion over their climate, Cactuses are in the most robust health, and their cells are abundantly filled with provision against scarcity. But these supplies would be quickly consumed by plants only protected by a thin cuticle, and having their surface pierced by millions of breathing pores, all actively exhaling the evaporable matter that lies beneath them, and an early death would be the inevitable result. Such, indeed, is the lot of all the gay companions of the Cactus, which surrounded it during the season of feasting and prosperity, and to which Nature has given no special means of enduring the hardships to which their lot exposes them; a few days or weeks suffice to sweep their forms from the face of the creation; their leaves rapidly consume the stores deposited in the stems, their stems turn in vain to the roots for a renewed supply, for after but a little while the arid earth has nothing to part with, and then the leaves wither and fall off, the stems shrink up and crack with the dry heat, and the roots themselves, in many cases, follow the same fate. With Cactuses this is different; they have so tough and thick a hide that what liquid substances they contain can only pass through it in minute quantities; the breathing pores of their surface are comparatively few, and so small as to act with extreme slowness when the air is dry; so that in proportion to the aridity of the air, and the heat to which such plants are exposed, is their reluctance to part with the food they contain. They digest and re-digest it, with extreme slowness, and may be truly said to live upon themselves during all those months when they cannot feed upon the soil or the atmosphere.

This statement applies more particularly to the species consisting of solid fleshy masses, like the Melon-thistles, the Hedgehog-thistles, and the like; but requires to be modified with reference to the thinner-stemmed species, such as Cactus speciosus, speciosissimus, and truncatus; of them it is equally true, but in a less degree.

The property which the Cactuses thus possess of living where few other plants can exist, sometimes renders them of great utility to man. On Mount Ætna, for instance, and its volcanic fields, it is the *Indian Fig* (Opuntia) which the Sicilians employ to render such desolate regions susceptible of cultivation. This plant readily strikes into the fissures of the lava, and soon, by extending the ramifications of its roots into every crevice of the stone, and bursting the largest blocks asunder by their gradual increase, makes it capable of being worked.

You will now be curious to know by what botanical characters these interesting plants are certainly known. To the tufted spines that are scattered over the stem, instead of leaves, I have already adverted. The flowers are the next part for us to study; and here you are at last introduced to the most highly developed, the most complicated, the largest, and the most richly coloured, or purely colourless, of all the blossoms in the Vegetable Kingdom. The *Showy Cactus* (Cereus speciosus) is at hand; by no means the handsomest or the largest of this glorious tribe, but one that shews as well as any other the nature of its organization (*Plate* XXX.). In the flower of this species, you will seek in vain for a distinction between the calyx and corolla. It has a cylindrical stalk (*fig.* 1.), the lower part of which (*a.*) is hollowed out for the ovary, and the upper portion covered with small scale-like rose-coloured bracts (*a. a.*), which gradually pass into large, thin, delicate leaves of the same colour, unfolding tier upon tier from within each other, and adhering by their lower ends, till a fleshy firm tube (*fig.* 4. *b.* & *fig.* 2. *a.*) is produced. About the middle of this tube, just where it swells out and ceases to be cylindrical (*fig.* 4. *c.*), springs forth a multitude of slender stamens (*fig.* 2. *b.*), placed row within row upon the tube, and forming a long, white, filamentous cylinder or cone. The ovary is, as you have already been told, a cavity in the bottom of the apparent stalk of the flower (*fig.* 4. *a.*); it contains a great number of young seeds, attached to the lining of the cavity, in eight rows, or placentæ, each hanging from the point of a long slender thread (*fig.* 5.). The style rises like a graceful column (*fig.* 4. *d.*), from the top of the ovary, and after reaching a little beyond the limb of the anthers, divides into eight, short, narrow, fringed arms, forming a beautiful star of eight rays. After a few days, or even hours, all this gorgeous panoply fades away, the stamens wither, the starry stigma closes its rays, and the style, no longer able to support it, curves downwards beneath its weight; the floral leaves droop, their colours become deadened, their firmness and elasticity are replaced by a soft and slimy ooze, and quickly afterwards the whole of this once lovely apparatus is thrown off by the ovary, which enlarges, becomes pulpy, acquires a new colour, matures its small brown seeds, and finally becomes a fruit so similar to that of a Gooseberry, that for a long time the latter and the Cactus were thought to be related. Its seeds contain an embryo (*fig.* 6. & 7.) coiled up in the shell, which accurately fits it, and having a long slender radicle, with two distinct cotyledons. This kind of structure, however, is not universal in the Cactus Tribe. It sometimes happens that the embryo is straight, and almost destitute of cotyledons, their presence being only indicated by a little notch in the end of the embryo (*fig.* 10. 11.). This unusual circumstance is interesting, as shewing that the habit of growing without leaves is not confined to the stem, but is to be met with, in some species, even in the embryo itself.

I have said that the fruit of the Cactus bears a strong resemblance to a Gooseberry; the similarity is not confined to the appearance, but extends to the flavour, texture, and quality. So wholesome, indeed, is the Cactus fruit, that it is an important object of cultivation in some countries. On Ætna, for example, the large cooling fruits of the Indian Fig are sold in considerable quantity, and some of the varieties are found of great excellence. In the West Indies, and South America, Cactus fruit is often consumed as Gooseberries.

Perhaps there are few plants more resplendently beautiful than the *Showy Cactus*, when covered, as it often is, with hundreds of its large rosy blossoms. But there are many species far more magnificent in their individual flowers; as for instance, all those called *night-blowing Cereuses* (C. grandiflorus, triangularis, Lanceanus, Napoleonis), with their large trumpet-shaped tubes, cut at the border into starry segments of the most dazzling white, the purity of which is increased by the tassel of pale yellow stamens that occupies their centre, and also by the extraordinary contrast of the beautiful flowers, and the misshapen, dingy, snake-like, leafless stems from which they spring. Many of the *Porcupine thistles* too, especially Echinocactus Eyriesii, partake of the same noble features; and as they have the property of flowering by day, they are the more valued as well as better known.

These particulars will make you as familiar with the Botanical history of Cactuses, as you perhaps already are with their general properties.

It may seem like a paradox at once to proceed from such plants as these, to Melons, Gourds, and Cucumbers, because of their natural affinity, especially if *Bryony* (Bryonia dioica, *Plate* XXX. 2.) be taken by way of illustration. And yet such is the course I must follow; for I know of no plants allied to Cactuses in so many respects as the GOURD Tribe is. This will be more evident presently.

That the various kinds of *Gourd*, *Vegetable Marrow*, *Squash*, and the *Melon*, *Water Melon*, and *Cucumber*, are all combined by characters of the strictest resemblance, requires no proof. Nor indeed is it possible to doubt that the Bryony (*Plate* XXX. 2.) also appertains to the smae group. I shall leave you to examine the former without my assistance; the last mentioned plant deserves a detailed notice. You are, perhaps, aware that it is a perennial plant, with a large fleshy poisonous root, and rough stems, that rapidly extend over bushes and hedges, adhering firmly to the branches by means of its tough aud numerous tendrils. In Norfolk, Suffolk, and many other parts of England, it is abundant in hedge-rows, half smothering the bushes it clings to, and reddening all the lanes with its clusters of scarlet berries.

It bears the rough, pale yellow, toothed leaves of the Gourd, but they are differently lobed and formed, for they have about five deep divisions, of which that in the middle is rather longer than those at the sides, while the lowest are often two-lobed, and always turned back upon the stalk, so as to give the leaf what is called a heart-shaped base. The flowers are in the technical terms of Botanists called diœcious; that is to say, those which contain the ovary and stigma grow on one plant, and those with the stamens grow on another plant. I must speak to you of these two separately.

The flowers with stamens (*Plate* XXX. 2. *fig.* 1.) have a green cupped calyx, with five little teeth (*fig.* 1. *a. a.*), and a light-green strongly veined corolla of five petals, forming part with the calyx so completely, that the whole has the appearance of one five-lobed calyx. The stamens are five in number (*fig.* 4.), they have no filament, but consist of a fleshy, lobed, or sinuous connective (*fig.* 4. *b.*), bordered by the narrow pollen-bearing cells of the anther, which are separated from the connective by a glittering row of little prominent glands, placed like a fairy necklace. Ovary there is none.

The flowers with a pistil, so far as the calyx and corolla are concerned, are like those containing the stamens, only smaller, and in closer clusters, with shorter stalks (*fig.* 2.). They do not contain a trace of stamens, but have an inferior, dark green, round, ovary (*fig.* 6. *a.*), ending in a short, stiff, round style, divided into three cushion-shaped stigmas (*fig.* 6. *d.*). When opened, the ovary contains some ovules, attached in double rows to three parietal placentæ (see p. 39), and is nearly filled up by a firm fleshy substance (*fig.* 7.). The fruit becomes a round, scarlet, pulpy berry (*fig.* 3.), containing two or three flat, brown, hard seeds (*fig.* 8. 9.).

If you compare what has now been described with

the structure of a Gourd, you will find that the principal differences are as follows. The Gourd has larger leaves and flowers, the latter being yellow; the sterile and fertile flowers both grow on the same plant; the anthers adhere together a little, and stand parallel with each other; the stigmas are two-lobed; and the fruit is a large seed-vessel, pulpy inside, but having a hard rind externally, and containing a great multitude of seeds. And if you examine others o the plants already named, you will see that the differences are of a similar description.

The most curious plants of the Gourd Tribe are the *Bottle Gourd* (Cucurbita lagenaria), which is fashioned like a flask, and the inside being removed is actually used as a water bottle, the *Snake Cucumber* (Momordica cylindrica), whose slender cucumber-like fruits are many feet long, and curved and twisted like a vegetable snake, and the *Spirting Cucumber* (Momordica Elaterium), the seeds of which are ejected with violence when the fruit-stalk is suddenly removed.

You will now sɛy, "I perceive the resemblance between all the plants you have named to me, and I understand their structure, but how do you show an affinity between the Gourd Tribe and the Cactus Tribe?" That is the next point.

In the first place, remember that the flowers of Cactuses are not always large and manifold in structure, but sometimes very small, and the parts far from numerous; secondly, that, as I have long since said (Vol. I. p. 105), the succulent character of Cactuses is not peculiar, but common to them with many others, and is hardly a mark of affinity, but rather a specific quality; thirdly, that many Cactuses are climbing plants, although they have no tendrils. These points being settled, remark in the next place, that both Cactuses and Gourds have succulent fruit; that their seeds are numerous, and attached to the sides of the fruit; that they have no albumen; and that there is hardly more difference between the calyx and corolla of the one than of the other; that is to say, that they are in both cases very similar to each other in appearance; moreover, that in each tribe the stamens grow from the sides of the calyx-tube, and the ovary is inferior. These resemblances are sufficient to show that the two tribes are allied to each other in no very distant degree, although they do not prove them to stand in immediate contact. But I have not asserted that such was the case; in fact, the most direct affinity of the Gourd is perhaps with the Passion-flower Tribe, as has been stated on a former occasion (Vol. I. p. 71.). From those plants, however, the Gourd Tribe deviates in many important particulars, so that, in reality, there is no known natural assemblage that they immediately impinge upon.

EXPLANATION OF PLATE XXX.

I. The Cactus Tribe.—1. A flower of Cereus speciosus, the natural size; *a a* the bracts; *b* the ovary.—2. The stamens, magnified, with a portion of the tube of the flower at *a*; *b* the filaments; *c* the starry stigma.—3. An anther, with a portion of a filament adhering to it.—4. A section of a part of the tube of a flower, with the ovary at *a*, the tube at *b*, the insertion of the stamens at *c*, and the base of the style at *d.* —5. A transverse section of the ovary, very much magnified, shewing the parietal placentation.—6. A seed of an Opuntia.—7. The embryo of the same.—8. Ripe fruit of a Mammillaria.—9. A seed.—10. A section of the same.—11. An embryo, with a notch at the end dividing the two cotyledons.

II. THE GOURD TRIBE. —1. A stamen-bearing twig of *Bryony* (Bryonia dioica).—2. A pistil-bearing twig of the same.—3. The ripe fruit.—4. A portion of the cup of a stamen-bearing flower, magnified; *a* the cup; *b* a single stamen; *c* a double stamen.—5. A bird's eye view of the lower part of a stamen-bearing flower, with a single anther at *a*, and two double ones at *bb*.—6. A portion of a pistil-bearing flower; *b* calyx; *c* corolla; *d* style; *a* ovary.—7. A transverse section of the ovary.—8. A seed.—9. A section of the latter, with the embryo.

LETTER XXXI.

THE BEGONIA TRIBE — THE FIG-MARIGOLD TRIBE — HYGROMETRICAL PHENOMENA CONNECTED WITH THE DISPERSION OF SEEDS.

Plate XXXI.

THERE are few collections in which some one or other of the plants called Begonias are not found. They are not, however, cultivated so much for the sake of their flowers, as of their leaves, the deep rich colours of which, especially their crimson, is unrivalled in the vegetable world. These plants have in all cases one half of the leaf much smaller than the other, so that at their base they often have something the appearance of a human ear. They have a pair of large stipules at the foot of each petiole, and all the parts of vegetation are particularly tender and brittle. They grow naturally in damp tropical woods, often on rocks, or in the rifts of trees, and are among the most certain signs of a hot damp climate.

It is a matter of no little difficulty to know where to class them, or with what plants they are most naturally related; indeed, after all the consideration that Botanists have given them, the subject is still unsettled. Why this is so, you will understand, as soon as I have explained to you the structure of the fructification of Begonia.

Let the subject of examination be the commonest

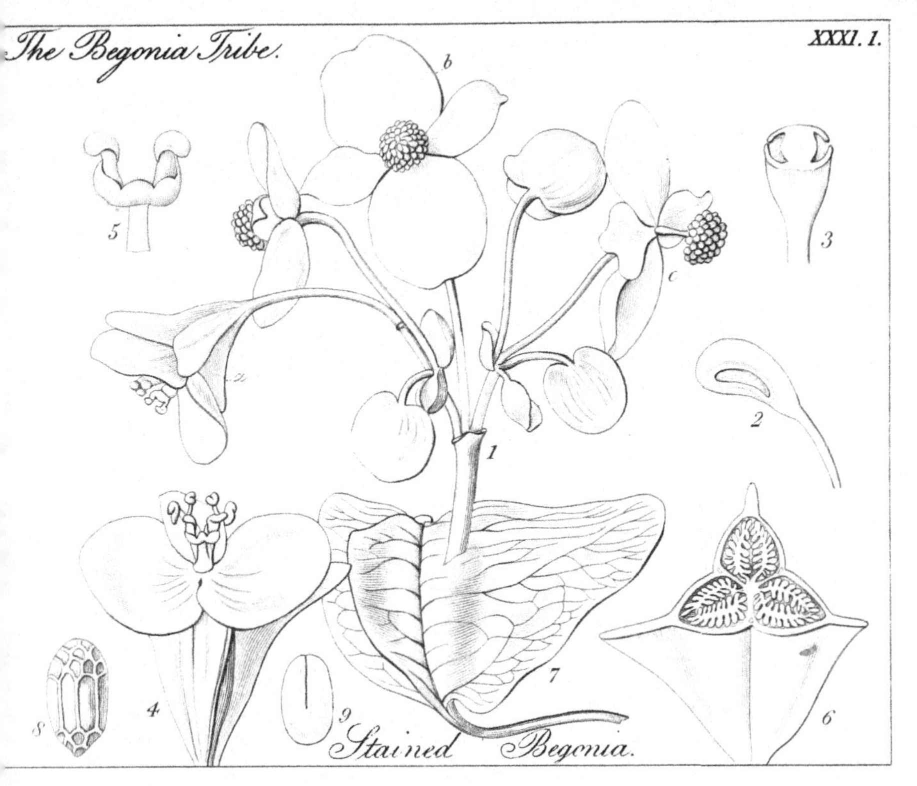
The Begonia Tribe.
XXXI. 1.
b
5
3
c
a
2
1
4
7
8
9
6
Stained Begonia.

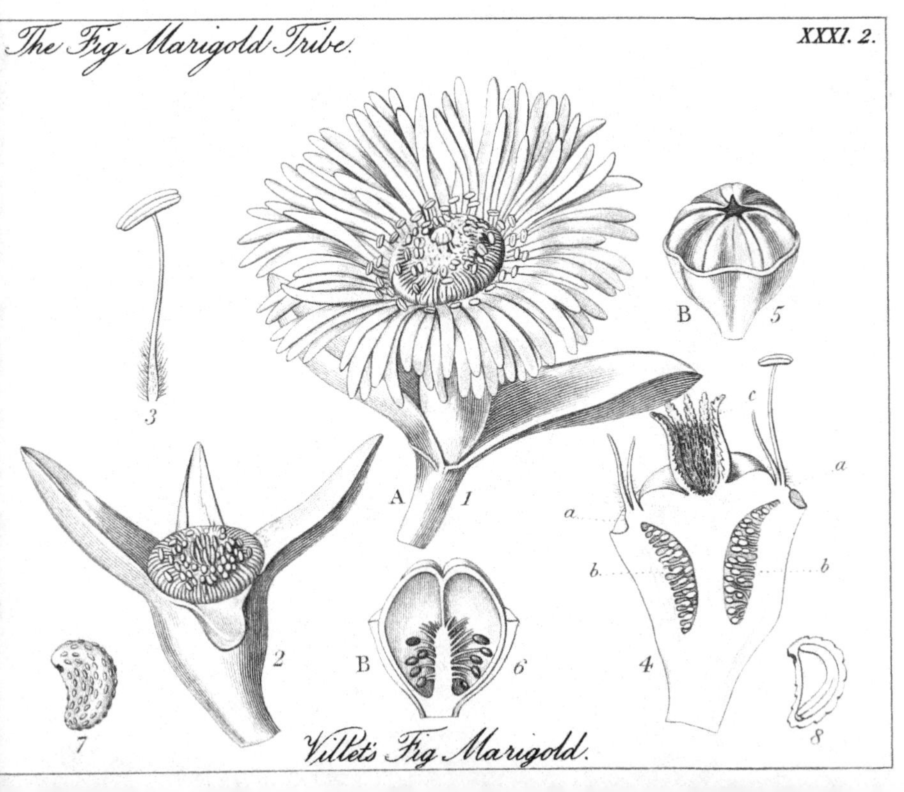
The Fig Marigold Tribe.
XXXI. 2.
B
5
3
c
a
A
1
a
b.
b
2
B
6
4
7
8
Villet's Fig Marigold.

of all the species, the *two-coloured Begonia* (B. discolor, *Plate* XXXI. 1.). The flowers of this plant grow in a kind of cyme, at the ends of the branches; each of the ramifications of the cyme has a pair of concave stipules at the base (*fig.* 1.); the flowers stand upon slender flesh-coloured stalks, and are partly sterile, partly fertile.

The sterile flowers consist of two larger obtuse pink sepals (*fig.* 1. *b.*), and two petals of the same colour. In their centre you have a round ball of anthers, the filaments of which are united into a common stalk (*fig.* 1. *c.*). The anthers are club-shaped, fleshy, yellow bodies (*fig.* 2.), having a curved pollen-cell on each side.

The fertile flowers (*fig.* 1. *a.* & *fig.* 4.) have a calyx and corolla like those of the others, only the latter has frequently but one petal. Beneath the calyx is a fleshy thick part, having three unequal wings (*fig.* 4.), divided into three cells internally (*fig.* 6.), with two plates or placentæ, covered with minute ovules, in each cell. Of course this part is the ovary; it is terminated by three stigmas, each of which (*fig.* 5.) has two twisted hairy lobes.

The fruit, when ripe, is a thin brown case, beautifully marked with deeper coloured veins, and having three wings, of which one is very much larger than the others (*fig.* 7.). It contains a multitude of small seeds (*fig.* 8.) of an oblong form, and covered with a network, the meshes of which are disposed with wonderful regularity; those at the two ends being always contracted and small, while the intermediate ones are long, with parallel sides; so that if a slice were cut off the lower end, the remainder would have quite the appearance of a gothic church window. The embryo, which lies in the seed, is an oblong succulent mass, half split into two parts (*fig.* 9.).

In attempting to fix the natural relationship of Begonia to other plants, we need not occupy ourselves with the little resemblances it may bear to this group, or that, in one or two particulars. But, as should be done in all such cases, I will beg you to confine your attention to its more striking peculiarities, and to their conformity with what can be found elsewhere. Now what are its more striking peculiarities? They may be collected under several heads; the stamens and pistils are in different flowers; the stigmas are two-lobed; the stamens are all combined into a single column; and the anthers have a remarkably thick connective; the calyx and corolla are in twos; that is, there are two sepals and two petals; and the ovary is inferior, three-celled, with many-seeded double placentæ.

Many groups of plants can be found, in which some one of these circumstances equally exists, but it is only when two at least occur, that a comparison can be usefully instituted. For example, the Cactus Tribe has a many-seeded inferior ovary; the Myrtle Tribe, in many cases, a three-celled inferior ovary; the Mallow Tribe, the stamens combined into a column; the Maple Tribe, a winged fruit; and so on; but in all these cases the resemblance can scarcely be traced further.

The natural assemblages in which the greatest number of points of resemblance can be found with Begonias, are the Euphorbia Tribe, to be examined hereafter, the Gourd Tribe, the Evening Primrose

Tribe, and the Buckwheat Tribe. For facility of comparison, we will make a little table, in which the most remarkable characters of these natural orders shall be placed side by side with what exists in Begonia.

Begonia Tribe.	Euphorbia Tribe.	Gourd Tribe.	Evening Primrose Tribe.	Buckwheat Tribe.
Leaves alternate	Yes.	Yes.	Sometimes.	Yes.
Large membranous stipules	Yes.	No.	No.	Yes.
Stamens & pistils in different flowers	Yes.	Yes.	No.	No.
Stigmas two-lobed	Sometimes.	Sometimes.	No.	No.
Stigmas combined in a column	Sometimes.	Sometimes.	No.	No.
Anthers with a very thick connective	No.	Yes.	No.	No.
Calyx and corolla in twos	No.	No.	Sometimes.	No.
Ovary inferior	No.	Yes.	Yes.	No.
Ovary three-winged	No.	No.	No.	Yes.
Ovary three-celled	Yes.	Three Placentæ.	No.	No.
Double placentæ	No.	Yes.	No.	No.
Seeds numerous	No.	Yes.	Yes.	No.
Albumen altogether wanting	No.	Yes.	Yes.	No.
Points of agreement	6	10	5	3
Points of disagreement	7	3	8	10

This shews you that it is to the Gourd Tribe that Begonias have the nearest relation: corresponding in ten important characters out of thirteen, and that of the orders thus brought into view, the weakest affinity is with the Buckwheat Tribe, or only as three to thirteen, and of those three characters, two are of the lowest importance. Indeed, I should not have thought it worth including the latter in the comparison, if it had not been the opinion of the learned Jussieu, that Begonias and the Buckwheats are related.

While, however, after an investigation of this nature, it is difficult to refuse assent to the placing Begonias and the Gourds near each other in the system, it is nevertheless obvious enough, that they are not so closely allied, as to deserve being considered contiguous groups; and it is highly probable that plants have still to be discovered, of an intermediate character, by means of which the two assemblages will be connected.

Before I dismiss the subject of Cactuses, and the orders allied to them, it is necessary that I should say a few words upon the FIG-MARIGOLD Tribe, an assemblage of plants of remarkable beauty, although but little cultivated now, in consequence of the fashion for Cape plants having gone by. The Tribe is represented by a genus called Mesembryanthemum, consisting of two or three hundred species, and to this my remarks will be confined. The principal part of the genus Fig-Marigold, or Mesembryanthemum, consists of shrubs inhabiting rocks and dry plains in the most arid parts of the southern extremity of Africa; they have fleshy leaves, often of most singular forms, and partake very much of that power of enduring drought, which, as you have seen, is one of the striking characters of Cactuses. Some of their leaves have a cylindrical form, and are terminated by a short tuft of bristles; in others, the leaves are curved like a Turkish scymitar, or fashioned like an axe; in some, they are rounded, so as to look like green pebbles collected into masses; and in several they are bordered by long stiff teeth-like fringes, and curve together so as to resemble the half-open jaw of some savage animal, whence the strange names of Tiger-chap, Dog-chap, Wolf-chap, Mouse-chap, and so on, by which different species are distinguished. Moreover, in one species, not a Cape plant, but an inhabitant of the North of Africa, the whole surface of the leaves and stems is raised into minute transparent blisters, so that the plant has the appearance of one of those beautiful French preserved fruits, which glitter all over with crystals of sugar; this species is known in the gardens by the appropriate name of *Ice-plant* (Mesembryanthemum crystallinum).

With regard to the fructification of this Tribe, it matters little what species we select. Here is one called in the gardens Villet's Fig-Marigold, nearly allied to M. acinaciforme, or the *Scymitar-leaved* (*Plate* XXXI. 2.). It has a succulent calyx of four or five unequal sepals (*fig.* 2.). Its petals are long, narrow, numerous, bright rose-colour, and closely packed one over the other in several rows (*fig. A.* 1.).

The stamens are numerous, and much shorter than the petals; they originate on the outside of a roundish, flat, green cushion (*fig.* 4. *a.*), that surrounds the stigma, and caps the ovary. The latter is inferior, containing about eight cells, divided off from each other by strong dissepiments, but, what is very remarkable, not bearing the ovules at the point where the dissepiments come in contact, but producing them from the centre of the back of each cell (*fig.* 4. *b.*). Hence in this species we have the singular instance of a many-celled ovary, with true complete dissepiments, and common parietal placentation. The stigma is sessile, and divided into as many rays as there are cells in the ovary. I must now warn you, that, although the species before us has this curious arrangement of the interior, yet you will not find the same structure in all species; on the contrary, in some, the back of the cell simply presents a fleshy hump, from the lower edge of which, and the base of the cell, the ovules originate; or, as in most cases, they simply grow from the lower part of the inner edge of the cell.

The latter structure is that of the ripe fruit I send you for examination (*fig.* B. 5.). You will find that it divides at the top into five valves, which close up when the fruit is wet, and open when dry. Each of its cells contains a considerable number of seeds (*fig.* B. 6.), hanging from long stalks, that grow from the lower part of the centre of the fruit. The seeds are angular, and tuberculated (*fig.* 7.), and contain a curved embryo, lying on one side of the albumen.

I have omitted to state, that in this and all the species, the flowers close in the shade, or in dull weather, and only expand under bright sunshine. I scarcely know a more interesting sight than in a summer's day, after a storm, to watch a bush of this genus, which has thrown its weak trailing arms over a piece of rock, and which leans forward to the south, as if to catch the earliest influence of the beams it loves so well. While the sky is darkened by clouds, all its blossoms are shut up so closely, that one would hardly suspect the bush of being more than a tuft of leafy branches, with some withered or unexpanded blossoms scattered over them. But the moment that the bright rays of the sun begin to play upon the flowers, the scene changes visibly beneath the eye; the petals slowly part, and unfold their shining surfaces, of almost metallic brilliancy, to the sunbeams, and in a few minutes become so many living stars, often of the most gorgeous tints, and so entirely hide the leaves, that scarcely a trace of them is visible, while the whole bush has burst into a blaze of glittering splendour.

In this case, the phenomenon depends upon a specific irritability of the petals, the cause of which is one of those inscrutable mysteries that the limited faculties of man are incapable of penetrating. But in the fruit there is an interesting phenomenon of another kind, the cause of which is more easily explained. The seed-vessels of the Fig-Marigold, produced, as I have just told you, in the sandy deserts of Southern Africa, fall off when ripe, and are driven about by the wind. If they were to open during the wet season, or in wet places, the seeds would fall out and perish, for it is only in a dry soil that they are capable of vegetating. Nature, therefore, gives this plant the power, by virtue of its hygrometrical quality, of keeping the seed-vessel fast shut up while exposed to damp, and it is only when it finds itself in a dry station, fit for the dissemination of the seeds, that the valves contract and open sufficiently to allow the latter to escape. It is impossible to imagine a more obvious interposition of Provi dence than this is, for securing the preservation of the race of the Mesembryanthemums.

But it is only one out of hundreds, that might be adduced to show the evident design that is visible in this part of the creation ; and, what is not less curious than interesting, where it is necessary for plants to disperse their seeds in the damp, nature provides for this also, with the most admirable certainty, by giving the valves of the seed-vessel the power of opening in humidity ; and so employing the same kind of power, that of hygrometrical action, for two opposite purposes. Thus, to use the words of the learned M. De Candolle, the Evening Primroses open the valves of their pods in wet weather, and close them when dry. This circumstance is probably connected with the manner of life of these plants, which naturally flourish in swampy places, and require to sow their seeds when the weather is wet. This notion is confirmed by the history of another plant having the same property, namely, that singular Eastern herb, known under the strange name of *Rose of Jericho* (Anastatica hierochuntica). This grows in the most arid deserts. At the end of its life, and in consequence of drought, its texture becomes almost woody, its branches curve up into a sort of ball, the valves of its pods are closed, and the plant holds to the soil by nothing but a root without fibres. In this state, the wind, always so powerful on plains of sand, tears up the dry ball, and rolls it upon the desert. If in the course of its violent transmission the ball is thrown upon a pool of water, then humidity is promptly absorbed by the woody tissue, the branches unfold, and the seed-vessels open; the seeds, which, if they had been dropped upon the dry sand, would never have germinated, sow themselves naturally in the moist soil, where they are sure to develope, and the young brood to be nourished. And in this way, a plant, to which the most silly superstition has given celebrity, really presents a truly marvellous phenomenon in its organization. Specimens of this curious production are sometimes brought from Palestine, where it is called Kaf Maryam, and, although they may be many years old, will, if placed in water, start, as it were, from their slumbers, stretch out their arms, straighten their leaves, and assume all the appearance of plants suddenly raised from the dead.

With regard to the affinities of the Fig-Marigold Tribe, it is obvious that generally they are with all the assemblages having both petals and sepals, many stamens, and an inferior ovary; such, for instance, as the Myrtle Tribe, and the Cactus Tribe; but it is especially with the latter that its consanguinity is most near; and it is not a little remarkable, that in the manner in which its fruit is constructed, and the ovules developed, it combines in some cases, in the same species, as we have seen, two different forms of placentation: the central and the parietal.

EXPLANATION OF PLATE XXXI.

I. The Begonia Tribe.—1. The inflorescence of the *stained Begonia* (B. discolor); *a* a fertile flower; *b c* sterile flowers.—2. A side view of an anther, with the cleft through which the pollen escapes. —3. A transverse section of the same.—4. A fertile flower.—5. One of the twisted two-lobed stigmas.—6. A transverse section of an ovary, shewing the three cells, in each of which there is a double placenta covered with ovules.—7. A ripe seed-vessel.—8. A seed very much magnified.—9. The embryo.

II. The Fig-Marigold Tribe.—A. 1. A flower of *Villet's Fig-Marigold* (Mesembryanthemum Villeti of the Gardens).—2. Its calyx and stamens.—3. A stamen.—4. A longitudinal section of the ovary; *a*, the insertion of the stamens; *b*, the parietal placentæ; *c*, the stigma.—B. 5. A ripe fruit of Mesembryanthemum, after Gærtner. —6. A longitudinal section of it, shewing the manner in which the seeds are attached to the bottom of the inner angle of the cells.—7. A seed.—8. A section of it with the embryo and albumen.

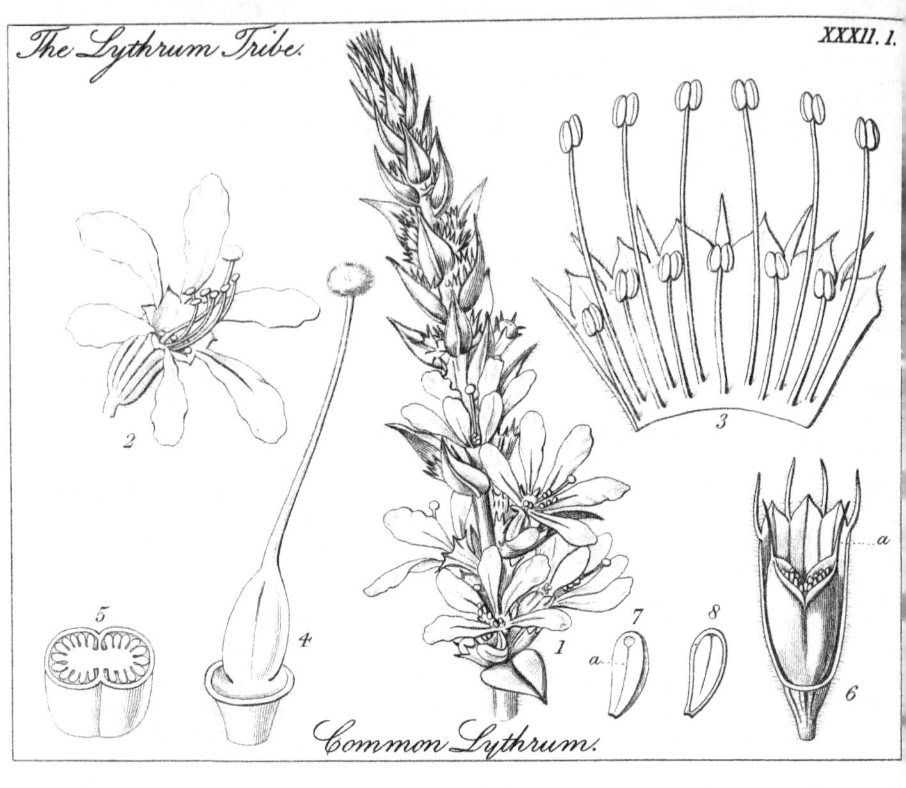
The Lythrum Tribe.
XXXII. 1.
2
3
a
5
4
1
7
a
8
6
Common Lythrum.

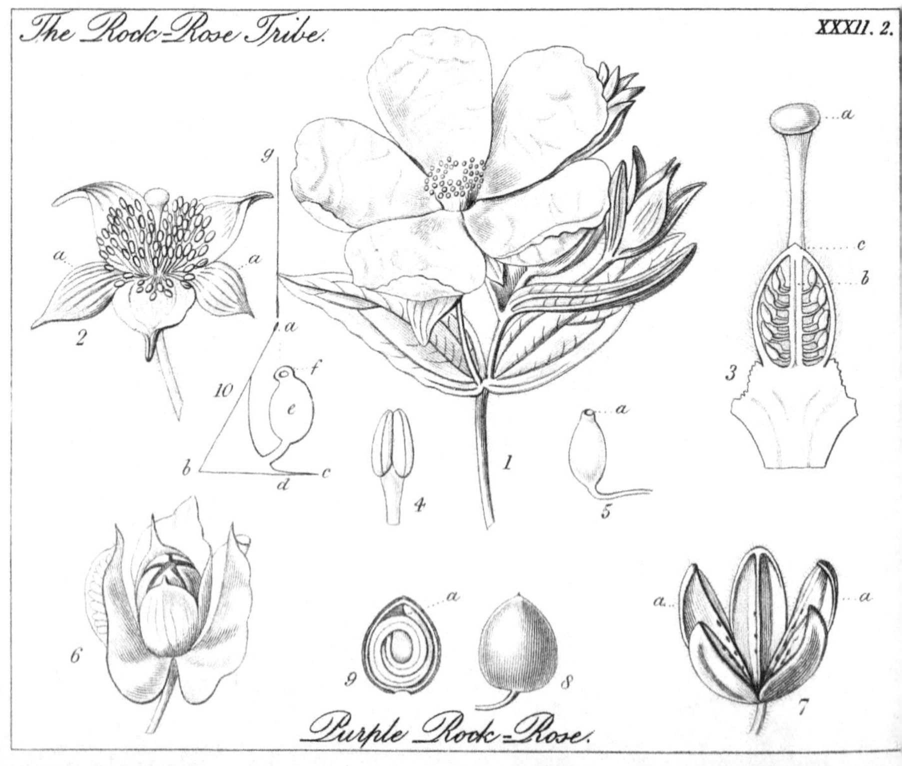
The Rock-Rose Tribe.
XXXII. 2.
g
a
a
2
a
f
10
e
b
d
c
4
1
a
5
a
c
b
3
6
a
9
8
a
a
7
Purple Rock-Rose.

LETTER XXXII.

THE LYTHRUM TRIBE — THE ROCK-ROSE TRIBE— MODE IN WHICH THE CONTENTS OF THE POLLEN-GRAINS ARE CONVEYED TO THE OVULE.

(*Plate XXXII.*)

In marshes, meadows, by the side of ditches, and, generally, in wet places, there grows a flower which, if it were brought from a distant country, reared in a hothouse, cultivated with difficulty, and sold at a great price, would be the pride of a collector, and the admiration of the crowd that is ever searching for new objects of amusement; for, proudly raising above the neighbouring grasses its long leafy rods, loaded with purple flowers, it stands confessed the undisputed queen of the meadows. But *Lythrum*—for such is its name—is only a wild flower; it may be had any where in autumn for the gathering; it associates with the sedge, and the meadow-grass, and ignoble weeds, and so, it is neglected, except by the few—are they indeed the few?—who love beauty for its own sake, and prize our fair native wild flowers, as much as costly strangers, which are only to be reared by wealth and skill, and which often owe their charms to the adventitious circumstances that surround them.

This plant has a hairy four-cornered stem, about

four feet high, rather closely covered with opposite lance-shaped leaves, which are always more or less hairy, and occasionally, even hoary. As the leaves approach the upper end of the stem, they become smaller, and at length form in their axils two or three flowers, of the following structure. The calyx is tubular, and pale green, with a red border ; it has twelve strongly marked streaks, or veins, traversing it in a nearly parallel direction, and it is divided at the edge into twelve little teeth, six of which are short and broad, and six much narrower and longer (*Plate* XXXII. 1. *fig.* 2. & 3.) The petals are six in number, narrow, blunt, crumpled, and light purple (*fig.* 2.). Twelve stamens spring from near the bottom of the calyx, in two rows ; one row is shorter than the calyx, the other much longer (*fig.* 3.), and both are curved towards one side of the flower. The ovary (*fig.* 4.) is superior, and has two cells, in each of which are many minute seeds, covering a central placenta (*fig.* 5.) ; the style is slender, and a little longer than the stamens, in the direction of which it is curved ; the stigma is a round velvety little cushion. When the fruit is ripe, it is closely covered by the dried calyx (*fig.* 6.), and is a capsule of two cells opening at the end, and bearing on each valve one half of the style (*fig.* 6. *a.*). The seeds (*fig.* 7.) are plano-convex, sharp-pointed at the base, and contain an embryo without albumen (*fig.* 8.).

This structure is remarkable in many respects ; in the first place, the striated calyx, and the square stem, both unusual circumstances, are analogous to what we find in the Labiate Tribe, which resembles the Lythrums in little else; then the seeds, the opposite leaves, the stigma, and the habit, are like those of some species of the Evening-Primrose Tribe, which differs, however, in having an inferior ovary, and in several more points; to the Mallow Tribe the Lythrums approach in their tubular calyx, crumpled petals, superior, many-seeded ovary, and double row of sepals; but their distinct stamens growing from the sides of the calyx, not to speak of other differences, prohibit the union of Lythrums with that order.

I will not fatigue you with further inquiries of this nature, but leave you to institute what comparisons you please between Lythrums and such natural groups as you are acquainted with. The result will necessarily be that they are not to be identified with any. Hence, Botanists class them in a distinct set, to which the names of Lythraceæ, Salicarieæ, or the LYTHRUM TRIBE, are given. The great features of the assemblage depend upon the plants being, 1. polypetalous, 2. having a furrowed or striated tubular calyx, 3. having stamens attached to the sides of the calyx, 4. a superior ovary, and, 5. numerous seeds without albumen.

In the gardens we have no common plants belonging to this order, except species of the genus Lythrum; but among the rarer plants are some that deserve mention.

In the first place, the Henna or Alkanna, with which oriental ladies stain their nails and the tips of their fingers a yellowish red colour,

> The Henna that is deeply dyed to make
> The skin relieved appear more fairly fair,

is obtained from a bush belonging to the Lythrum tribe; for this purpose its leaves are pounded, and made into a paste. Botanists call the plant Lawsonia inermis; antiquaries have asserted, without much reason, that it is the Gopher plant of Scripture.

A second object of interest is the beautiful timber used by Cabinet-makers under the name of Rosewood. By some this production is assigned to a plant called Physocalymna floribunda belonging to the tribe before us; but Prince Maximilian of Wied Neuwied declares that it is yielded by a Mimosa.

While speaking of the remarkable plants of the Lythrum Tribe, the Lagerströmias must on no account be forgotten: Indian and Chinese trees or shrubs, bearing a profusion of large purple flowers, in clusters of considerable size, and one of them (L. indica), at least, nearly hardy in England.

The Rock-Rose Tribe *(Plate* XXXII. 2.) shall be the next object of our examination, and most worthy of it will it prove, whether the beauty of the species belonging to it, or their very extraordinary structure be considered. These plants are well known in gardens, under the names of Cistus or Helianthemum, and are either cultivated as evergreen bushes in the shrubbery, or are employed to ornament rough banks and masses of rock-work, over which they trail or spread with great beauty; they are particularly useful in places so much exposed to the sun as to be too dry in summer for the support of other plants. In such situations they grow with vigour, resist severe frosts, and all the summer long are every morning adorned with an inconceivable profusion of night-born blossoms, which drink in with avidity the first rays of the sun, but, after a few hours, perish beneath his fervid rays. The colours of these blossoms are yellow, or yellow spotted with deep brown, purple, rose-colour, white spotted with purple, or the most pure and dazzling white. The leaves, moreover, of the Cistuses give out a delicious balsamic odour, which, in places where the plants are numerous, literally fills the air, especially after a shower, with a slight, but most agreeable and reviving fragrance. In their native countries, particularly in the south of France, Spain, and the Islands of the Mediterranean, the Cistuses are by far the most lovely objects that Nature has planted in the woods, rocks, and other stations they inhabit.

In their foliage they are not sufficiently uniform for the leaves to form a part of their distinctive character, which in this instance is derived principally from the fructification. The *purple Rock-Rose* (Cistus purpureus) will give you a good example of it.

In that species you have a calyx composed of five pieces (*fig.* 2.), which, however, do not exactly form a single row or whorl; but, as you may see by tearing them off, two (*fig.* 2. *a. a.*) grow a very little lower down than the three others, which, moreover, are something larger and a little paler at the edges; such a calyx is said to form a *broken whorl.* The corolla (*fig.* 1.) consists of five equal purple petals, which, from the manner in which they are packed up within the bud, have a crumpled appearance when the flower unfolds. A great many stamens, much shorter than the petals, grow in a ring from below the ovary (*fig.* 2.). The ovary itself (*fig.* 3.) is superior, with five cells, in each of which are many ovules, rising upwards upon slender curved stalks, and pointing towards the top of the cell. Each ovule is egg-shaped, and has a perforation, called a *foramen*, at its point (*fig.* 5. *a.*). The style is taper, and rather thicker at the upper than the lower end; the stigma (*fig.* 3.) is a convex undivided space, abruptly terminating the style, and bordered by a delicate fringe of hairs.

When the seed-vessel of this plant is ripe, it is enclosed within the calyx, grown larger, harder, and deep brown (*fig.* 6.). It consists (*fig.* 7.) of five boat-shaped valves (*a. a.*), along the middle of each of which passes a ridge that was, in the ovary, a dissepiment, and to which the numerous seeds adhere. The seeds are little, smooth, stalked, heart-shaped bodies (*fig.* 8.), pointed at the upper end, and containing an embryo, coiled up in the most curiously careful manner (*fig.* 9.); the embryo itself is imbedded in a small quantity of albumen, and, contrary to what usually occurs in other plants, the radicle is placed next the point of the seed (*fig* 9. *a.*).

Such are not only the characters of the Purple Rock-Rose, but also in a great measure of the whole tribe. The common genera differ from each other, chiefly in little points, that in no way interfere with the more striking features; such for example, as having only three sepals instead of five, having the seed-vessel very imperfectly divided into cells by short partitions, and so on.

It must be obvious to you, when you come to consider the resemblance of the Rock-Rose Tribe to others, that it has a strongly marked analogy with Poppies (Vol. I. *plate* 1. *p.* 19.). They both have crumpled petals, which fall off soon after they expand, a great many stamens growing beneath the ovary, an ovary with parietal placentæ, and numerous seeds. But, on the other hand, they are separated by many equally remarkable differences, as you will see by the following contrast.

Poppy Tribe.	Rock-Rose Tribe.
Parts of flower 3 or 4.	Parts of flower 5.
Calyx in a perfect whorl, and soon falling off.	Calyx in a broken whorl, and remaining on the plant as a protection to the seed-vessel.
Ovules with the foramen next the base.	Ovules with the foramen at the point.
Embryo straight and very minute, in a large quantity of albumen.	Embryo rolled up, filling the inside of the seed, almost to the exclusion of the albumen.
Radicle of the embryo next the base of the seed.	Radicle of the embryo at the point of the seed.

I have just mentioned that the Rock-Rose Tribe has a very extraordinary structure; let me now explain in what that consists. You have already remarked that the ovule (*fig.* 5.) has a perforation or foramen at its point; all ovules have such a perforation, but not all in the same place. In most ovules it is next the base, in a few only does it exist at the point, as in the plants before you. The use of the foramen is not a little curious. You are aware that when the ovule is first formed it is no more than a mass of pulp, in which little or no organization can be detected internally; but in process of time a small cloudy speck forms in this pulpy interior, and keeps growing larger and larger, till at last it becomes an embryo. It has been observed that the speck always first becomes visible next the foramen; and there is great reason to believe that in reality the speck is introduced into the ovule through the foramen. Further, it is supposed that it is in the anther that this speck is first formed; that it originates in the inside of a grain of pollen; that when the pollen falls upon the stigma, the former puts forth an excessively fine tube, much finer than the most delicate hair; that the tube passes down the style, and continues to lengthen till it reaches the foramen; that the contents of the grain of pollen are discharged into the tube, and the speck with them; that it is then, by some hidden and mysterious agency, carried down the tube; and that, finally, it is thus conveyed into the ovule through the foramen. For all the evidence, and the many curious facts, connected with this part of botany, I must refer you to modern Introductions to the subject; in this place, you must be satisfied with my assurance, that this extraordinary statement is supported, not only by observations of my own, but by the concurrent testimony of all the most cautious and skilful microscopical observers who have engaged in the inquiry.

What I have already stated to you is extraordinary enough, and much cause as you have already found at every step to admire the wonderful care and skill with which all the actions of vegetable life are conducted, yet I think you must here find a fresh and unexpected source of admiration. You see, that in the formation of the seed of even what we may deem the most worthless weed, there is the same unerring Providence, as in the preservation of the race of man. Only think for a moment, upon the long long journey that the little speck, the tiny rudiment of a seed, has to take before it can arrive at the only place in which it is possible that its destiny can be fulfilled, or that it should be developed into a new being. Born in the pollen-grain, it is originally enclosed in a doubly guarded prison: its own little spherical vault, and the more extensive walls of the anther. The anther must open before the pollen can escape; and it must open too at a particular time, at the very moment when the stigma has secreted a clammy dew, which will hold fast the pollen if it falls upon it. Then the pollen must fall on the stigma; to fall elsewhere is useless. This accomplished, the microscopic rudiment of the seed, which, although not exactly an *être de raison*, for it can be discovered with the microscope, is practically so to human eyes—this almost invisible particle, has to commence a long and winding journey through all the intricacies of the style, and the ovary, till its guardian tube conducts it to the ovule and deposits it in safety. And all this is so provided for, that we find every adjustment exactly that which is best suited to the object in view; invisible springs in the anther, acted upon by the very same cause as that which renders the stigma clammy, combine their million little forces to pull open the sides of that case; to enable their forces to act with certainty, the sides of the anther are weakened in a particular line, which in every anther of the same species is constantly the same. It is supposed that the clamminess of the stigma is not merely to stick the pollen-grain fast, but also to cause the formation of the pollen-tube; to enable the latter to reach the ovule, notwithstanding its excessive delicacy, the whole texture of the stigma and style is loosened, so as to offer as little resistance as possible to the passage of the pollen-tube. In this Rock-Rose Tribe we have a still further example of the facility with which obstacles to communication between the pollen-tube and the opening in the ovule are overcome.

If we suppose a grain of pollen to fall on the stigma of a Cistus (*fig.* 3. *a.*), its tube may be easily understood to reach the place where the ovules grow (*fig.* 3. *b.*); but, when there, it is cut off from the foramen by the whole length of the stalk and sides of the ovule, for the foramen is at the other end of the latter. In order to overcome this difficulty, we are told by M. Adolphe Brongniart, that the pollen-tube does not follow the placenta till it reaches the ovule (at *b.*), but quits the style at the top of the cavity of each cell (*c.*), and thence lengthens in the open space inside the ovary, in the form of the finest imaginable cobweb, till it reaches the foramen in the end of the ovules.

To make this clearer, observe the following diagram (*fig.* 10.). Let the perpendicular *a. g.* represent the style, the line *a. b.* the side of the ovary, the horizontal line *b. c.* the base of the ovary, the curve *a. d.*

the placenta, *e.* the ovule, and *f.* its foramen; then the pollen-tubes may be stated to quit the style at *a.*, to hang down freely in the cavity of the ovary, in the direction of the dotted line *a. f.*, and thus to secure a short line of communication with the foramen.

Many more such cases are to be found by those who search for them; but none much more curious than the present.

EXPLANATION OF PLATE XXXII.

I. THE LYTHRUM TRIBE.—1. A twig of Lythrum Salicaria, the *Purple Loosestrife*, in flower.—2. A flower slightly magnified.—3. The calyx cut open, showing the two rows of stamens, and the manner in which they adhere to the calyx.—4. An ovary, with the style and stigma.—5. A horizontal view of the interior of the ovary, showing the ovules adhering to the placentæ.—6. Half a calyx, with the ripe fruit in the inside; *a* one of the halves of the style, carried away on the point of the valve of the fruit when it burst.—7. A ripe seed; *a* the raphe, with the chalaza at its end.—8. A longitudinal section of the seed, showing the dicotyledonous embryo.

II. THE ROCK-ROSE TRIBE.—1. A flower and leaves of the *Purple Rock-Rose* (Cistus purpureus).—2. A calyx with the stamens and ovary, *a a* the two outer sepals.—3. A longitudinal section of the ovary; *a* the stigma, *b* the placenta; this gives a good view of the ovules.—4. An anther.—5. An ovule; *a* the foramen.—6 A ripe seed-vessel, invested with the calyx.—7. A seed vessel burst; the seeds fallen out; *a a* valves.—8. A seed.—9. The same cut longitudinally, showing the embryo rolled up, with the radicle at *a.*—10. A diagram to explain the manner in which the pollen-tubes reach the foramen of the ovule.

LETTER XXXIII.

THE TAMARISK TRIBE—THE SUN-DEW TRIBE—HAIRS OF PLANTS.

(Plate XXXIII.)

WE have scarcely a prettier shrub in our gardens than the Tamarisk, with its long, deep-brown, slender rods, delicately studded near the points with green scale-like leaves, or bowing beneath the weight of graceful plumes of faintly blushing blossoms; in their native places the species are still more striking. On the sea-beaten cliffs of a wild shore, the dry rocky bed of a winter torrent, the naked plains of Egypt, the islands of the Nile, the wilderness of Sinai, and the desolate coast of the Red Sea; in these and other such places the Tamarisk rises with its greatest grace and beauty.

There is something in the habit of this plant so peculiar, that the Botanist has always been puzzled to determine with what others it should be allied; and after one incongruous association or another, it seems now settled that it has no very marked affinity with other plants, but really possesses so peculiar a structure as to form a little group by itself.

In the gardens are two distinct kinds of Tamarisk, one called the *French*, with dark chocolate-brown branches (Tamarix gallica), and the other called the

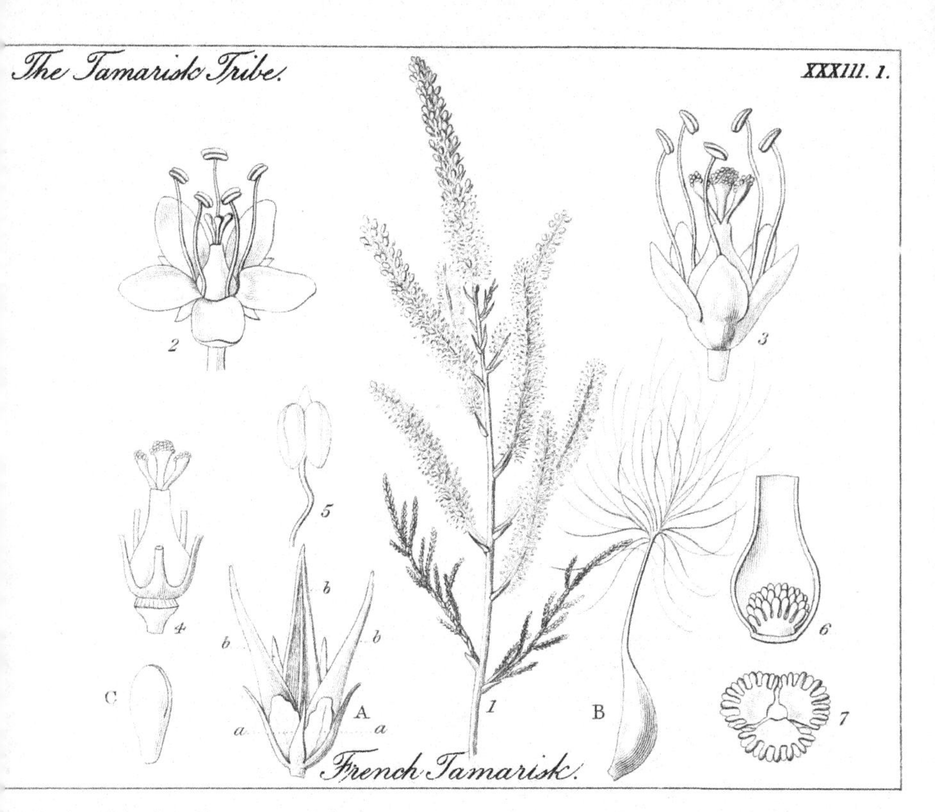
The Tamarisk Tribe.
XXXIII. 1.
2
3
4
5
6
7
1
b
b
b
a
a
A
B
C
French Tamarisk.

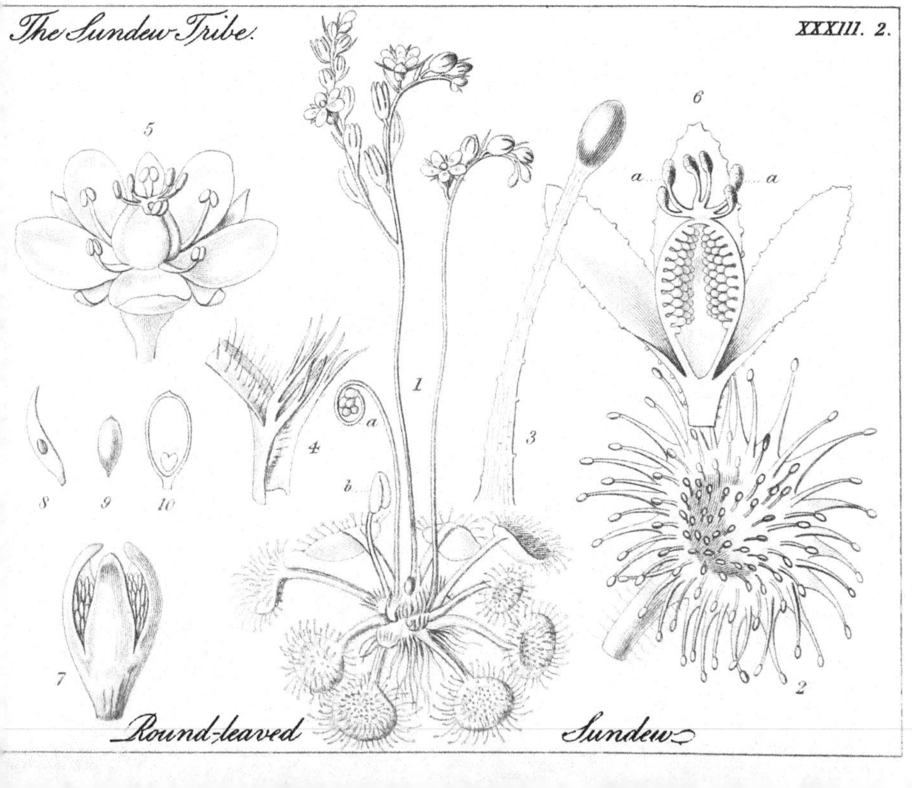
The Sundew Tribe.
XXXIII. 2.
5
6
a
a
1
a
b
3
4
8
9
10
7
2
Round-leaved
Sundew.

German, with a sea-green bark (Myricaria germanica). The former is much the handsomer of the two, and the one we will take for examination.

Its leaves are little green scales, closely packed one above the other, and covering the stem while it is young, but quickly falling off as the branches grow old. The same peculiarity is found in all the plants of the TAMARISK TRIBE. So is the manner in which the flowers are placed, in long, graceful, tail-like racemes, at the extremity of the branches (*Plate* XXXIII. 1. *fig.* 1.).

Each flower consists of a sea-green calyx, having a cup-like downy tube, and five oval lobes delicately bordered with pink (*fig.* 3.); of five spreading white petals (*fig.* 2.); five stamens growing from below the pistil (*fig.* 4.); and a superior ovary. The latter is pale yellow tinged with pink (*fig.* 4.), shaped something like a flask, and suddenly ends in three white styles, each terminated by a thick granulated stigma. The interior of the ovary consists of a single cavity (*fig.* 6.), at the very bottom of which lie three convex placentas covered with ovules (*fig.* 7.).

I do not find this kind of Tamarisk with ripe seed-vessels, but here is that of the German Tamarisk (*fig. A.*) which does as well. It is divided into three valves, each of which has an elevated ridge along its middle, and is surrounded at the base, not only by the dried up calyx, but also by the petals which adhere to the seed-vessel in the form of little scales (*fig. A. a.*). The seeds in this species are terminated by a long beak, the end of which is surmounted by a tuft of hairs (*fig. B.*), doubtless intended to enable the seed to ride on the wind, and to be transported from place to place; in the French Tamarisk this provision exists only in a very rudimentary state. In the inside of the seed lies an embryo with two cotyledons, and no albumen (*fig C.*).

What renders the French Tamarisk still more interesting than its graceful form, is the belief that it was from this plant, or a local variety of it, that the manna fell, on which the Israelites subsisted during their sojourn of forty years in the deserts of Arabia. The celebrated Professor Ehrenberg gathered manna with his own hands from the Tamarisks of the wilderness of Sinai, and it is certain that the species grows plentifully in all the countries adjacent to the Red Sea. That manna did fall from the Tamarisk, is rendered more probable by the fact that this substance is at the present day produced by only two plants in the East, one the Tamarisk, the other the *Camel's Thorn* (Alhagi Maurorum); but as the manna of the Mosaic history is said to have fallen from heaven, that is, from some height, it could scarcely have been produced by the Camel's Thorn, which is only a low bush, while it might easily have dropped from the Tamarisk, which becomes a tree. It is, moreover, not a little curious that the Tamarisk manna is very different in its effects from the bitter sweet manna of the druggists' shops, which is sometimes given to infants as medicine; Tamarisk manna is stated by the chemists, who have examined it, to consist of pure mucilaginous sugar, one of the most nutritious of known substances.

You will presently see that so far as you have any means of judging upon such points, the Tamarisk Tribe has a near resemblance to the Sun-dews or Droseras in some respects, although the resemblances are in reality those of analogy only, and not of affinity.

Queen of the Marsh, imperial Drosera treads
Rush-fringed banks and moss-embroidered beds;
Redundant folds of glossy silk surround
Her slender waist, and trail upon the ground.
As with sweet grace her snowy neck she bows
A zone of diamonds trembles round her brows;
Bright shines the silver halo, as she turns;
And, as she steps, the living lustre burns.

It is thus that Dr. Darwin introduces one of the most curious little plants in the world; and although the exact rules of science will necessarily repudiate such language, yet it must be confessed that there is much poetical truth and beauty in the description.

You will, I am quite sure, be anxious to make acquaintance with Drosera, who would rather seem to be a fairy than a plant, by the poet's description; but I fear there is little chance of your beholding her upon her own moss-embroidered bed, unless by accident: for her home is the fen and the marsh, the oozy heath and the treacherous morass, where she takes possession of every little hillock elevated somewhat above the surrounding waters, and whence no art can induce her permanently to depart. If you snatch her from her native soil, and cherish her with the most curious care, you will hardly succeed in prolonging her existence beyond a few short months. Let her, however, be sought for by all means, and she will richly reward you for any trouble you may take in procuring her. When she is in your possession, plant her among some bog-moss in a saucer or deep dish, place over her a bell-glass, pour water into the dish till it rises above the rim of the glass, then expose her to the full rays of the sun, and you will have done all that art can effect to secure her.

The structure of Drosera is the following, if you take *the Round-leaved* (Drosera rotundifolia), which is the commonest species, as an example. In this plant the most remarkable part is the leaves, and the least remarkable the fructification. The former are nearly round, and grow upon long hairy stalks; they are at first folded up in such a manner that they look something like little green hoods (*Plate* XXXIII. 2. *fig.* 1. *a.*), but they afterwards spread out into small concave disks, covered over with long, shining, red hairs, that secrete from their point a clear fluid, which gives the leaves the appearance of being covered with dew-drops. Real dew is, you know, always dispersed and dried up by the heat of the sun, so that it is only at the earliest hours of the morning that it can be seen in the summer; but the glittering dew-like secretion with which the leaves of this plant are bespangled is most abundant when the sun is at his highest, and hence it has acquired its popular name of Sun-dew; as if the particles of water which cause the leaves to sparkle were really dew, condensed by the sun's rays.

The apparatus by means of which the moisture is

secreted, forms one of the most beautiful of objects for the microscope. Let us take a single hair, and place it under a magnifying glass, taking care to throw upon it from above a strong reflected light, and using the precaution of cutting off all the rays that come from below. You will now see that what seemed a little hair with a drop of water at its point, is really a long curved horn, transparent and glittering like glass; delicately studded from top to bottom with sparkling points; beautifully stained with bright green passing into pink, and mellowing into a pale yellow, as if emeralds, rubies, and topazes had been melted, and just run together without mixing; and finally tipped with a large polished oval carbuncle, or ruby of the deepest die (*fig.* 3.). In this there is no exaggeration; for what tints can possibly represent the brilliancy of vegetable colours, except those of the purest and noblest of precious stones?

If you observe this organ a little more carefully, you will remark a number of faint streaks running side by side from its lower to its upper end, and interrupted at brief but pretty regular intervals, by exceedingly short transverse lines. These marks are external indications of the cells that the organ is composed of; and if you take the trouble to compute the number of such cells required to form it, you will find that there must be at least two thousand such cells in each of these little horns. Every one of such cells is continually absorbing, and secreting, and digesting the fluids that pass into it from the leaf, or from the air; so that for the due performance of the office of such a minute body as a hair of the Sun-dew leaf, no fewer than two thousand little digesting cells, or stomachs, are incessantly exercising and combining their tiny forces!

There is still the ruby-coloured point to examine. In its interior structure it is like the hair itself, only all the parts are more solid; it is here that the fluid secreted by the hair is finally concentrated; and it is from this that the dew is continually exuding, so as to stand upon it like a drop of water. The water has a slightly acrid taste, and is probably thrown off from the leaf, because its continued presence in the system of the Sun-dew would be pernicious.

The hairs of our British Droseras possess the power of closing upon insects and holding them fast. "When an insect settles upon them, it is retained by the viscosity of the glands, and in a little while the hairs exhibit a considerable degree of irritability by curving inwards, and thus holding it secure."—(Henslow.) And Dr. Royle describes the phenomenon as occurring so obviously in an Indian species of Sun-dew, that he had called it "the fly-catching" in consequence.

The description just given of the hairs of the Sun-dew, is in part applicable to all other hairs; for they are generally constructed upon a similar plan, and are often, when filled with moisture, most beautiful and elaborately constructed organs. Botanists distinguish two principal sorts of hairs; the *glandular*, in which the hair is either tipped with a secreting organ, as in the Sun-dew, or arises from one, as in the Borage Tribe; and the *lymphatic*, in which there is no secreting organ present, beyond the cell or cells of the hair itself. For a particular account of them you must turn to works more explanatory of the structure of plants than these letters are intended to be.

Near the base of the leaf-stalk is a long coarse fringe (*fig.* 4.), which is supposed to represent stipules.

The flowers of the Sun-dew, when expanded, are elevated upon a slender scape, along one side of the upper end of which they are arranged; but when young, they are coiled up in a gyrate (or circinate) manner (*fig.* 1. *a.*). The calyx consists of five sepals, a little glandular externally, and nearly as long as the petals (*fig.* 5. and 6.). The petals are five, snow-white, flat, blunt, and spreading (*fig.* 5.). There are five stamens, growing from below the ovary, opposite the sepals. The ovary is a superior, oblong case, of one cell, and bears three clusters of ovules on its sides (*fig.* 6.); it is surmounted by three forked stigmas. The fruit (*fig.* 7.) is a capsule, half divided into three valves, and enclosing a multitude of minute seeds. Each seed (*fig.* 8.) is invested in a loose membranous tunic tapering to each end, and contains a kernel (*fig.* 9.) filled with a large quantity of albumen, in the base of which is a minute two-lobed embryo (*fig.* 10.).

Many as have been the differences in the combination of the floral organs, in the numerous tribes of plants already examined by you, this is manifestly one to be added to your list; for in no others have you hitherto met with the union of a coiled inflorescence, a few hypogynous stamens, parietal placentæ, and a minute embryo lying in the base of the albumen. These characters, independently of all others, distinctly separate the Sun-dews as a peculiar tribe. What the plants really are, to which they are most nearly related, is still an unsettled point. Violets, Saxifrages, Frankenias, have been respectively selected ; but there are objections to all those natural groups. It is probable that the true affinity of the Sun-dews is with Side-saddle Flowers, most curious plants inhabiting the marshes of North America.

EXPLANATION OF PLATE XXXIII.

I. THE TAMARISK TRIBE.—1. A twig of *French Tamarisk* (Tamarix gallica).—2. A flower, magnified.—3. A calyx seen in profile, with the stamens and ovary, the petals being removed.—4. An ovary, with the bases of the five stamens grown into a sort of cup, and surrounding it.—5. An anther, with a portion of the filament.—6. A section of the ovary, shewing how the ovules rise from a convexity in the bottom.—7. A section of the convexity (or placenta), shewing that it is not single but triple.—*A*. Ripe fruit of the German Tamarisk (Myricaria germanica) ; *a a* the withered petals ; *b b b* the valves of the capsule.—*B*. A seed of the same.—*C*. it sembryo.

II. THE SUN-DEW TRIBE.—1. A plant of *Round-leaved Sun-dew ; a* a young scape, rolled up in a circinate manner ; *b* a young leaf before expansion.—2. A magnified leaf, showing the glandular hairs.—3. A glandular hair, very highly magnified.—4. The lower end of a leaf-stalk, with the stipulary fringe.—5. A flower, magnified.—6. A section of the ovary, exhibiting the parietal placentation ; *a* a stigma.—7. A ripe capsule.—8. A seed, very highly magnified ;—9. its kernel.—10. The same divided lengthwise, and exhibiting the embryo at the base of fleshy albumen.

XXXIV.

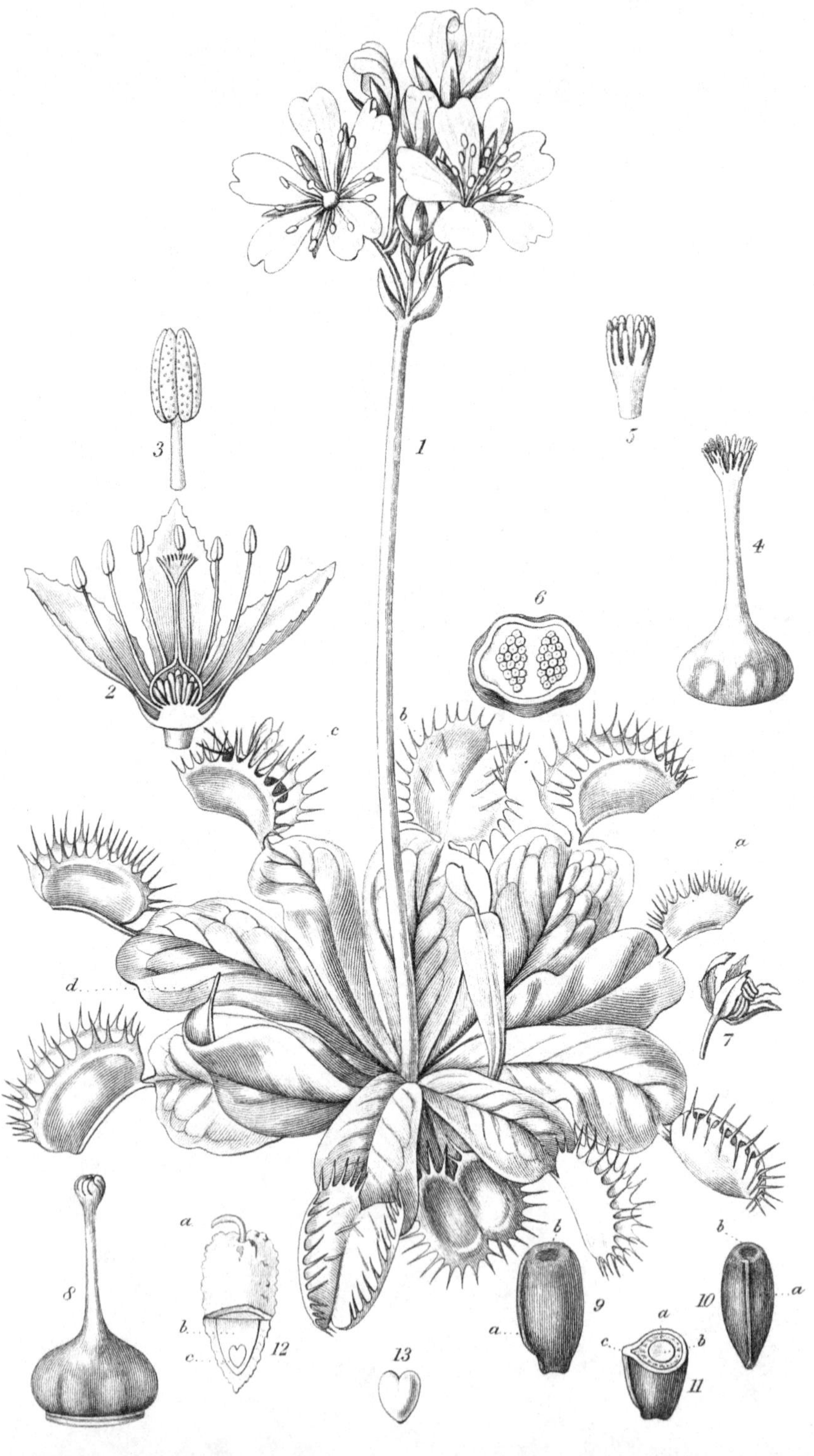

LETTER XXXIV.

VENUS'-FLY TRAP—ANATOMICAL STRUCTURE OF LEAVES.

(Plates XXXIV. and XXXV.)

ARE you acquainted with a most singular plant, the *Venus' Fly-trap* (Dionæa muscipula), an inhabitant of turfy and sandy bogs in the warmer parts of the United States?* If not, search for it immediately in

* I copy the following account of Dionæa, in its American home, from a work on the plants of North Carolina, by Mr. M. A. Curtis, as quoted in the Companion to the Botanical Magazine.

"The Dionæa muscipula is found as far north as Newbern, North Carolina, and from the mouth of Cape Fear River nearly to Fayetteville. Elliott says, on the authority of General Pinckney, that it grows along the lower branches of the Santee, in South Carolina, and I think it is not improbable that it inhabits the Savannahs more or less abundantly from the latter place to Newbern. It is found in great plenty for many miles around Wilmington in every direction.

"I venture a short notice of this interesting and curious plant, not being aware that any popular description of it has been published in this country. The leaf, which is the only remarkable part, springs from the root, spreading upon the ground, at a little elevation above it. It is composed of a petiole, or stem with broad margins, like the leaf of the orange tree, two to four inches long, which, at the end, suddenly expands into a thick and somewhat rigid leaf, the two sides of which are semi-circular, about two-thirds of an inch across, and fringed around their edges with somewhat rigid ciliæ, or long hairs, like eye-lashes. The leaf, indeed, may be very aptly compared to two upper eyelids, joined at the nurseries, place it among bog-moss in a green-house, and cover it with a bell-glass, keeping it constantly damp. In this manner you may secure for a year or two one of the most curious examples of irritability which the vegetable world contains; and

---

their bases. Each portion of the leaf is a little concave on the inner side, where are placed three delicate, hair-like organs, in such an order that an insect can hardly traverse it, without interfering with one of them, when the two sides suddenly collapse and enclose their prey, with a force surpassing an insect's attempts to escape. The fringe or hairs of the opposite sides interlace, like the fingers of the two hands clasped together. The sensitiveness resides only in these hair-like processes on the inside, as the leaf may be touched or pressed in any other part, without sensible effects. The little prisoner is not crushed and suddenly destroyed, as is sometimes supposed, for I have often liberated captive flies and spiders, which sped away as fast as fear or joy could hasten them. At other times, I have found them enveloped in a fluid of mucilaginous consistence, which seems to act as a solvent, the insects being more or less consumed by it. This circumstance has suggested the possibility of the insects being made subservient to the nourishment of the plant, through an apparatus of absorbent vessels in the leaves. But as I have not examined sufficiently to pronounce on the universality of this result, it will require further observation and experiment on the spot to ascertain its nature and importance.

" It is not to be supposed, however, that such food is necessary to the existence of the plant, though, like compost, it may increase its growth and vigour. But however obscure and uncertain may be the final purpose of such a singular organization, if it were a problem to construct a plant with reference to entrapping insects, I cannot conceive of a form and organization better adapted to secure that end, than are found in the Dionæa muscipula. I therefore deem it no credulous inference, that its leaves are constructed for that specific object, whether insects subserve the purpose of nourishment to the plant or not. It is no objection to this view, that they are subject to blind accident, and sometimes close upon straws, as wells as insects. It would be a curious

which, in some respects, is more striking than even the Sensitive plants themselves, for they merely shrink away from the touch, while this plant firmly grasps, with its wonderful leaves, anything that comes within their reach. Its near connection with the subject of the last letter induces me to dwell upon its peculiarities at some length, independently of its own most interesting organization.

Its leaves spread in a circle round the crown of the root, and either lie flat upon the ground, or gently elevate themselves above the soil. They have no stipules, or stipulary fringes, but consist of two parts, very distinctly separated from each other, and remarkably different in their nature; one of these parts is a stalk and the other a blade, but both so much disguised as hardly to be recognised. The stalk is a flat, green, wavy, obovate, very obtuse, leafy expansion, the veins in which are coarsely netted, with curved branches, which, growing to each other's backs, form a number of somewhat lozenge-shaped meshes (*Plate* XXXIV. 1.). The blade is joined to this by a very narrow neck, and consists of a roundish, thick, leathery plate, slightly notched at each end, having strong hidden parallel veins, which spread, at nearly a right angle,

---

vegetable, indeed, that had a faculty of distinguishing bodies, and recoiled at the touch of one, while it quietly submitted to violence from another. Such capricious sensitiveness is not a property of the vegetable kingdom. The spider's net is spread to ensnare flies, yet it catches whatever falls upon it; and the ant-lion is roused from his hiding-place by the fall of a pebble; so much are insects, also, subject to the blindness of accident."

from the midrib to the margin, and bordered with a row of strong, stiff, teeth-like hairs. When young, the two sides of the blade are placed face to face, and the teeth cross each other *(fig.* 1. *a.)*; afterwards, when full grown, the sides spread flat, or nearly so, and the teeth then form a firm spreading border *(fig.* 6.*)*. On each half of the blade, stand three delicate almost invisible bristles, uniformly arranged in a triangle. If one of those bristles is touched, the two sides collapse with considerable force, the marginal teeth crossing each other, so as to enclose securely any small object which may have caused the irritation, or pressing firmly upon the finger, when the irritation is produced by it; but wonderful to relate, no other part of the leaf is sensible to external impressions. It is in vain that the back of the leaf is disturbed, or that the smooth glandular surface of the face is irritated; unless you jar one of these bristles no irritability whatever is excited, and the leaf remains immoveably open. The moment the shock is communicated through one of the bristles, the collapse of the leaf is effected, which then assumes altogether the appearance of an iron rabbit-trap when it has closed upon its prey *(fig.* 1. *c.)*. If, at this time, an attempt is made to open the leaf, it is violently resisted, in consequence of the rigidity of the side veins, whose contraction seems to be connected with the phenomenon. Upon this subject I shall not dwell any further just now.

The flowers grow in a cyme at the top of a scape, six or seven inches high. They consist of a calyx of five tooth-letted sepals, five very blunt petals, slightly two-lobed at the point, ten stamens growing from beneath the pistil, and of a superior ovary (*fig.* 2.). The anthers are covered over with little glittering glands. The ovary has a depressed form, something like that of an old German wine-bottle (*fig.* 4.); it contains but one cell, in the very bottom of which are two flat placentæ (*fig.* 6.), from which a great number of ovules grow erect; it gradually tapers into a green column of a style, the point of which is split into a ring of fringes (*fig.* 4. & 5.), and forms a stigma. The seed-vessel is a small flask-shaped capsule (*fig.* 8.), closely covered over by the calyx, and remains of the corolla. It contains a considerable number of black, oblong seeds, that are discharged only after the decay of the seed-vessel, which has no means of spontaneously opening. The seeds have a conspicuous raphe (*fig.* 9. & 10. *a.*) and chalaza (*fig.* 9. & 10. *b.*), and contain a kernel enveloped in a soft spongy substance (*fig.* 11. *a.*). The kernel is principally composed of albumen, the embryo (*fig.* 12. *c.* & 13.) being a very small two-lobed body.

Upon comparing this with the structure of the Sundew, it must be obvious to you, that the number of points of identity is extremely numerous, and that, in reality, the most important differences consist in the number of stamens being greater in Dionæa, there being but two placentæ, and those arising from the base of the capsule, the seed-vessel not bursting, the seeds not having a loose integument, the stigma not having twice as many lobes as placentæ, and the leaves being destitute of stipulary fringes upon their stalks.

Such distinctions would be more important, if many more species, corresponding with one another in habit, were found to possess them; but as there is nothing in the habit of Dionæa, materially at variance with that of Sun-dew, and as only one species of the genus has ever been seen, it is not considered absolutely necessary to separate it from the Sun-dew Tribe; especially as the position of the placentæ at the base, instead of the sides of the seed-vessel, is not esteemed of any structural importance. Nevertheless, it is to be remarked, that the flower-cyme is not coiled up, in a circinate manner, before the flowers unfold, that there is no trace of a tendency in Dionæa, to open its seed-vessel by valves, and that the loose integument of the seed of Sun-dew has no parallel in Dionæa.

Such are the principal circumstances deserving notice in the fructification of the Venus' Fly-trap. Let us now recur to the highly curious phenomenon from which it derives its name. You have seen that the upper surface of the blade of its leaf is extremely irritable, so that, when it is touched never so gently, the two sides collapse forcibly; it has been said, that this irritability invariably resides in three bristles, similar to the teeth of the margin, but much finer, and growing from the surface of the leaf in a triangular order. Why it is, or by virtue of what power, the bristles possess the key to the irritability of the Dionæa leaf, no one has ever succeeded in discovering. The phenomenon seems to belong to the extensive class of final causes which man is not permitted to explain. We, moreover, find upon the surface, a prodigious multitude of red glands, so minute as to be individually invisible to the naked eye, but giving a red tinge to the leaf. Such glands are found nowhere except upon the upper surface of the leaf, in the neighbourhood of the delicate seat of irritability. It is in vain that you stimulate the teeth of the margin, the back of the blade, or its stalk; in none of these parts is there a trace of irritability; and in none of these parts is there a trace of the glands. It is not, therefore, improbable that these glands are either in some way connected with the irritability, although it is not they through which the shock is first communicated to the leaf, or, as Mr. Curtis supposes, are intended to absorb the nutriment afforded to the leaf by the decay of the insects entrapped in it.

Let us be a little more particular in the examination of the Dionæa leaf; for it will not only give you instruction in respect to the plant actually before you, but will afford an insight into the general nature of the anatomy of all leaves.

With an exceedingly sharp, thin-bladed knife, obtain a thin slice of a leaf, in the direction of its veins (*as from b to e in Plate* XXXIV. *fig.* 1.), so as to shew its whole thickness. Place it under a good microscope, in water, and by means of the mirror throw light upon the slice from below; it will then become a transparent object, and you will be able to see all that minute, internal organization, which is entirely invisible to the naked eye, and which enables the leaf to breathe, perspire, digest, and perform its other manifold offices. You will also find that a leaf is not a thin homogeneous mass of firm pulp, nor a confused mixture of pulp and fibre; but a most elaborate, and yet simple apparatus, in which every part is adjusted with the utmost nicety; that, moreover, thin as the leaf appears, it is actually composed of at least nineteen or twenty layers of cells, besides a large line of vessels in its middle. That you may understand this the better, let me refer you to the accompanying sketch of such a slice as I have been talking of (*Plate* XXXV. *fig.* 1.). Let *A* be the upper surface, and *B* the lower surface. The upper surface is protected by a very thin, transparent, rather tough, homogeneous membrane (*a.*), which overlies all the cuticle, except perhaps the stomates, and does not appear to be in any degree cellular. It is not improbable that a similar membrane is found on the upper side of all leaves; it has been seen in the Cabbage, the Foxglove, &c. but has not hitherto been much investigated. On the lower surface of Dionæa-leaf, this membrane is absent. Immediately beneath the membrane, comes the skin or cuticle (*b.*), which, although it may be stripped off, nevertheless consists of long, flat, thick-sided cells, adhering very firmly to each other. This you will see more distinctly, if you strip off a piece of the skin from another portion of the leaf, and place it in water, in the same manner (*figs.* 2 & 3.).

From the cuticle of the upper surface there spring, at very short intervals, little red glands (*fig.* 1. *d d d.*), which grow from minute, green, oval spaces, composed of two, parallel, green cells, and resembling stomates.

These are the glands already referred to. They are firm fleshy bodies, resembling little convex buttons; and are composed of cells, arranged in a circular manner, round an axis, consisting of two such cells, stationed one on the top of the other (*fig.* 4. & 5.). I presume that these glands are analogous to the curious hairs of Sun-dew, although we do not see that they are possessed of any irritability; but in the Sun-dew they arise from a general expansion of the cuticle, and not from spurious stomates.

The cuticle of the under-side of the leaf is similar to that of the upper; but it is destitute of glands (*fig.* 1. *e.* & *fig.* 3.), in lieu of which little clusters of transparent greenish hairs (*f.*) grow from the abortive stomates. These hairs are each composed of one single cell, and may be considered a rudimentary form of the glands of the upper surface of the leaf (*figs.* 6. & 7.). On the under surface, however, you will find, in addition, a considerable number of true stomates, or breathing pores (*fig.* 3. *a a a.*). What those organs are, and for what purpose they are believed to be intended, has already been explained to you (Vol. I. *p.* 103.).

Immediately beneath the cuticle of either surface of the leaf lies the parenchyma, or pulpy part (*fig.* 1. *c c.*), composed of several layers of cells, gradually growing larger, more transparent, and thinner-sided, as they approach the middle. The cells of parenchyma are supposed to be the principal seat of digestion and respiration. The food of the plant is propelled into the leaf through the woody tubes, next to be spoken of, from them it is given off to the parenchyma, where it is gradually changed by the complicated processes of digestion, and whence it is returned into the body of the plant. Below the parenchyma run the woody tubes or fibres (*fig.* 1. *g g.*), which are in this plant short cylinders, but which more generally are very long and flexible; they compose a sheath, 3 or 4 layers thick, to protect the spiral vessels (*h.*): highly elastic tubes, capable of unrolling in a spiral direction, and supposed to be connected with the respiration of plants.

In the accompanying sketch, all the cells of the parenchyma are represented as being in close contact with each other; but, in reality, there are many open spaces among the cells, arranged in no regular order, and believed to be intended for facilitating the passage of air from one part of the interior of a leaf to another.

It is far from being my intention to explain any further, in this place, the anatomical structure of leaves. That of Dionæa gives you a sufficiently just idea of the general plan on which they are formed internally; for more exact information, I must refer you to the higher elementary works on Botany.

EXPLANATION OF PLATES XXXIV. AND XXXV.

Plate XXXIV.—1. An entire plant of *Venus' Fly-trap* (Dionæa muscipula) in flower, and bearing leaves in different states; *a* represents a leaf before it is expanded; *b* is another fully open; *c* is a third which has closed upon an insect; *d* is a dilated leaf-stalk, on which the blade of the leaf is not formed.—2. A section of a flower, with the petals removed; exhibiting the origin of the stamens, the position of the ovules, and the form of the stigma.—3. An anther, and the upper end of the filament.—4. An ovary.—5. A stigma, closed after fertilization has taken place.—6. A bird's-eye view of the bottom of the inside of the ovary, with the two placentas.—7. A ripe seed-vessel, invested by the withered calyx and corolla; of the natural size.—8. A ripe seed-vessel magnified, with the calyx and corolla stripped off.—9. A seed seen from the side; *a* the raphe, *b* the chalaza.—10. A seed seen from the edge; *a* the raphe, *b* the chalaza.—11. A cross section of a seed; *a* the spongy substance (secundine?) lying between the testa and the nucleus, *b* the nucleus, *c* the raphe.—12. A kernel taken out of the testa; *a* a portion of the raphe, *b* the albumen, *c* the embryo.—13. An embryo.

Plate XXXV.—1. A highly magnified view of a slice of the leaf of Dionæa, taken in the direction of the veins; *A* the upper surface, *B* the under; *a* the outer integument; *b* and *e* the cuticle; *c c* the parenchyma; *d d d* the glands; *f* one of the tufts of hairs arising from an abortive stomate; *g g* the woody tubes that surround the spiral vessels; *h* a bundle of spiral vessels.—2. A bird's-eye view of the skin of the upper surface; *a* the outer integument, through which the cuticle is seen; *b* a gland; *c c* abortive stomates.—3. A bird's-eye view of the skin of the lower surface; *a a* perfect stomates; *b b* abortive stomates; *c* a tuft of hairs arising from an abortive stomate.—4. Bird's-eye view of a gland very highly magnified.—5. A side view of the same.—6. 7. Views of one of the tufts of hairs that grow upon the under surface.

LETTER XXXV.

THE HORSE-CHESNUT TRIBE—THE WALNUT TRIBE.

Plate XXXVI.

You must have often admired the Horse-chesnut tree, either when rising in solitary beauty on the broad greensward of a highly cultivated park, or when, in the form of an avenue, great numbers of those trees combine into high banks of deep green foliage, and gayly tinted flowers. Let us take this plant as our next subject of examination, for which purpose we will select the rose-coloured species (Æsculus rosea, or carnea, *Plate* XXXVI. 1.).

Its leaves, you see, are regularly opposite each other on the branches, and are divided into several toothed lobes, which all proceed from one common point at the top of a strong round foot-stalk. The flowers appear in compact, erect, stiff panicles, at the ends of the branches. Their bractes are small, and quickly wither away, falling off, and leaving a scar behind them. Their calyx is a fleshy, smooth, reddish cup, divided at the edge into five unequal, oblong, blunt lobes. The petals are four only; their claw is long and channelled, and inserted below a one-sided, wrinkled, inconspicuous disk (*fig.* 2. *a.*); their limb is oblong, crumpled, crisped, of a bright yellowish red colour, changing

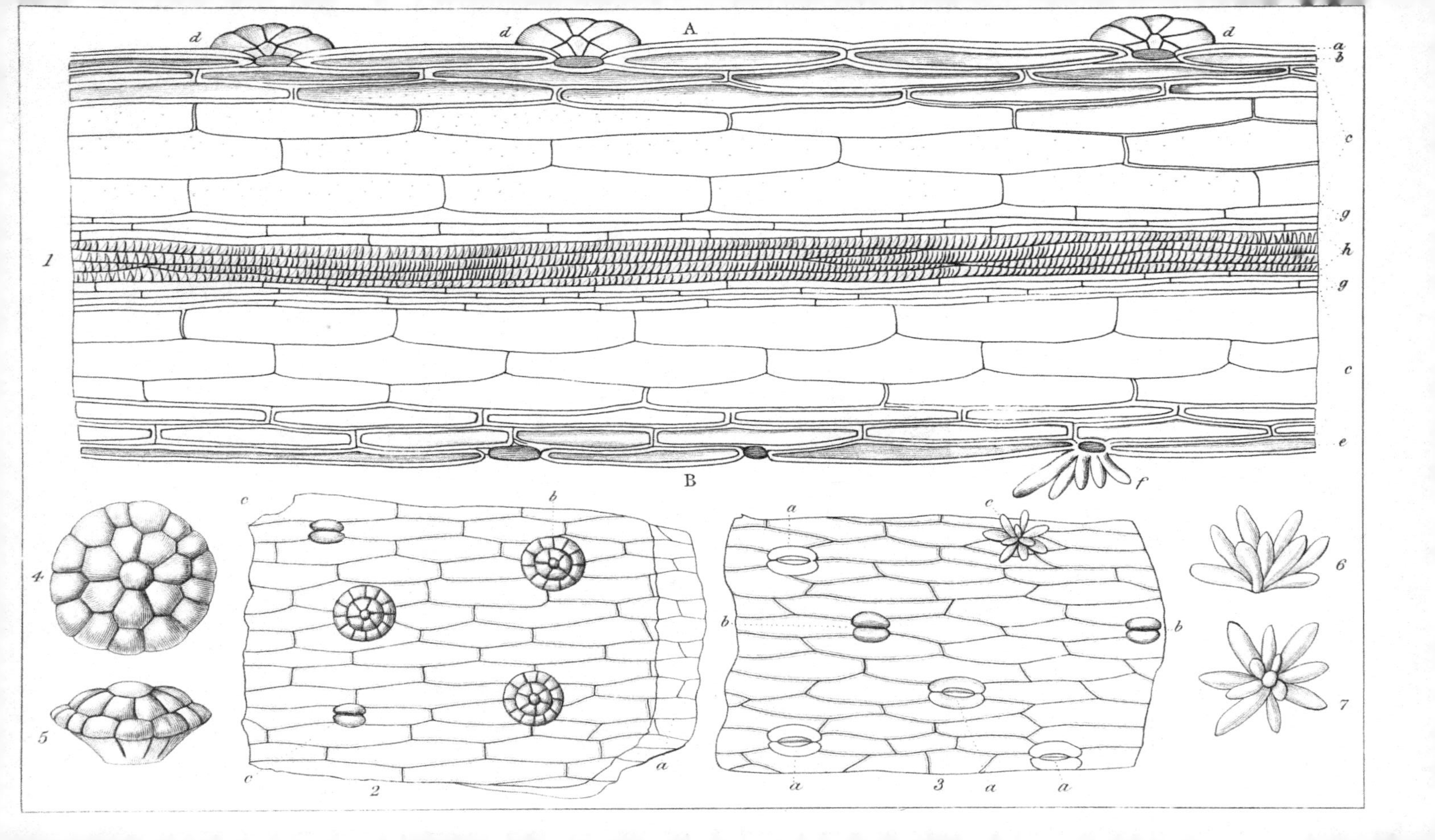
A
B
a
b
c
d
e
f
g
h
1
2
3
4
5
6
7

into bright orange-yellow at the base, and covered with soft hairs; two of the petals stand at the back of the flower, and two at its sides, overlapping the former a good deal, and exceeding them considerably in size; a fifth petal is wanting from the front, and hence this flower is both unequal and unsymmetrical in its corolla. This irregularity occurs also in every part, except the ovary. We have already seen that the lobes of the calyx are unequal; the disk has also been described as one-sided; and you will next find that the stamens are unsymmetrical, with regard to the surrounding parts. Instead of being five or ten, and so corresponding with the calyx, or four or eight, which would agree with the petals, you will find only seven, which symmetrizes with neither; they are curved downwards towards the front of the flower, their filaments are covered with long hairs (*fig.* 2.), which protect the style, and they terminate in oblong, red, hairy anthers, tipped with a reddish point (*fig.* 5.). The pistil is covered with hairs, and bent forwards and downwards in the direction of the stamens. It has a simple style, the point of which, where the stigma is, has no hairs, and a fleshy two or three celled ovary (*fig.* 3.), the sides of which are deeply channelled by the pressure of the filaments. In each cell you will find two ovules, one of which rises up, while the other hangs down, from a projecting horizontal placenta (*fig.* 4.).

The fruit of this plant becomes an unequal-sided, leathery, muricated seed-vessel (*fig.* 6.), opening by two or three valves, and containing one large roundish seed in each cell. The seeds (*fig.* 7.) have a hard, shining, deep-brown coat, a very broad scar (*fig.* 7. *a.*), on one side, and a little conical elevation, which touches with its point one edge of the scar (*fig.* 7. *b.*). This conical elevation represents the position of the radicle of the embryo that is hidden beneath the seed-coat. Let the latter be removed; you will find below it a roundish, wrinkled, fleshy body, which you cannot separate into cotyledons, but whose radicle, curved down upon itself, is distinctly visible. Here we have one of several instances, where the cotyledons grow to each other, so as not to be separable. The plumule, or growing point, of this embryo lies closely packed between the bases of the consolidated cotyledons, and one wonders how it is to escape from them, when the time shall arrive for the seed to commence its growth into a plant. A simple alteration in the adjustment of the parts produces the desired effect. As the cotyledons cannot unfold in the usual manner, in order to allow the plumule to pass between them, the passage of the latter upwards into the air is provided for by a slight extension of the bases of the cotyledons, which begin to lengthen when the radicle forces itself into the earth, and thus extricate the plumule from what would otherwise be its prison-house.

The structure that exists in the Red and the Common Horse-chesnuts is nearly the same as what occurs in the other species of the order, which is an extremely small one. The Pavias, or Scarlet Horse-chesnuts, are the only others that deserve notice, and they are so conformable in structure as not to require separate mention. Indeed, the order itself is chiefly introduced into this correspondence, by way of illustrating points to be hereafter adverted to.

A much more uncommon structure than that of the Horse-chesnut is found in the Walnut, with some account of which, as it is so very common and useful a tree, I may as well fill up the remainder of this letter.

Although my observations will be confined to the common Walnut, they will also apply to the principal part of the WALNUT TRIBE, in which are comprehended all the nuts named by the Americans of the United States, *Hickories*, and from which the Red Indian makes his bows.

The common Walnut is, as you know, a tree of very large size, producing valuable timber, and having, when old, a most majestic appearance; hence Botanists have named it the *Kingly Walnut* (Juglans regia). It bears long pinnated leaves, something like those of the Ash, but placed alternately upon the stem, and having, when bruised, a strong balsamic odour.

The chief peculiarity of the Walnut consists in the fructification, which, while it approaches in some respects that of the Oak Tribe (Vol. I. *p.* 138.), is of an essentially different and very peculiar nature.

The stamen-bearing flowers are on one part of the branch, and the pistil-bearing on another, as in the Oak and its allies. The former (*Plate* XXXVI. 2. *fig.* 1.) are arranged in thick, green, curved, cylindrical spikes, consisting of very short pedicels (*fig.* 2.), bearing obliquely on one side about twelve stalkless broad anthers, surrounded by about six green scales. These spikes fall off soon after the anthers have burst and discharged their pollen.

The pistil-bearing flowers, grow in clusters of two, three, or more (*fig.* 4.), and are composed of an oval, downy ovary, crowned by a minute four-lobed calyx (*fig.* 4. *a.*), four very small petals (*fig.* 4. *b.*), and a pair of fringed stigmas, curved in opposite directions. The interior of the ovary presents a minute cavity, in which is one erect, egg-shaped ovule (*fig.* 5. *a.*), seated on a pale lobed substance, a longitudinal section of which is extremely similar in form to the Russian eagle. The latter substance may be supposed to contribute to the nutrition of the embryo, but its use has not been yet sufficiently inquired into.

In course of time, the stamen-bearing flowers fall off, as has already been stated, the pistil-bearing flowers alter their appearance, lose their stigmas and all trace of a calyx and petals, become much increased in size, and at last change to clusters of oblong, deep-green, fleshy cases (*fig.* 7.), which crack irregularly and drop, leaving behind them the pale brown tesselated nuts, that are sold in the fruiterers' shops (*fig.* 8.). Examine one of these nuts, with which you ought to be well acquainted, because it is of such every day occurrence; and you will find that it might serve as a text for a long and curious disquisition. With only the most striking points however do I propose to occupy your attention.

The nut of the Walnut Tree, deprived of its outer fleshy shell, is of the same nature as the stone of a Peach or Plum; that is to say, it is the innermost layer of the seed-vessel, grown very hard, and separating from the outer layer. At a very early period (as for instance in the state of *fig. 5.*), the two layers formed but one homogeneous body; and when the inside began to harden, without any corresponding change in the outside, still the two held firmly together by a network of veins, the impressions of which give rise to the channels that divide the surface of the nut into numerous irregular compartments.

In one respect the nut of the Walnut differs essentially from the stone of a Peach. In the latter it is not divisible into valves; in the former it readily separates into two equal valves. These are an evidence, although only one ovule is present, yet that this fruit is in reality made up of two carpels, as was indicated by its two recurved stigmas. Now examine the valves separately; each is cut off from the other at the base, by an imperfect partition that rises up from the very bottom; but, above the base, they freely communicate with each other. Their inner surface is marked by numerous elevations and hollows, of a most irregular arrangement, besides which a small plate, originating in the partition at the base, but standing at right angles to it, curves upwards, and cuts each valve imperfectly into two cells; so that, what with the partition at the base, and the plates at right angles with it, the interior of the nut is, before it is opened, cut into four incomplete cells.*

In the centre, where these imperfect plates cross each other, stands the seed, which in growing adapts itself both to the plates themselves, and to the inequalities in the lining of the nut, so that when full grown it is four-lobed, and deeply divided all over by irregular fissures (*fig.* 6.).

The seed, like the ovule, stands erect in the cavity of the nut; but the embryo is inverted, its base or radicle (*fig.* 6. *a.*) being at the point of the seed. The cotyledons are applied face to face, and each participates in the convolutions of the other, until they meet the elevated point of the central plate on which the seed rests; thence they separate in a downward direction, and consequently each pair of shrivelled seed-lobes consists of one cotyledon only.

* In technical language this nut must be described as consisting of two opposite connate carpels, whose margins at the base are turned inwards towards the placenta, whence they are partially produced as far as the back of the cavity of the carpel, forming an adhesion with it, and half dividing the cavity into two spurious cells.

EXPLANATION OF PLATE XXXVI.

I. The Horse-chesnut Tribe.—1. A panicle of flowers of the *Pink Horse-chesnut* (Æsculus rosea).—2. The stamens and disk (*a*) of one of the flowers.—3. The ovary, with the one-sided disk (*a*) at its base.—4. A longitudinal section of the ovary, shewing the ovules in their two different positions.—5. A stamen.—6. A seed-vessel, natural size.—7. A seed; *a* the scar or hilum; *b* the conical projection on one side of the scar, indicating the position of the radicle.

II. The Walnut Tribe.—1. A portion of a twig of the *Common Walnut* (Juglans regia), with a stamen-bearing catkin.—2. One of the stamen-bearing flowers, in the position in which it hangs in the catkin.—3. A stamen.—4. Two pistil-bearing flowers; *a* the calyx, *b* the petals.—5. A longitudinal section of one of these flowers; *a* the ovule, *b* the calyx, *c* one of the petals.—6. A ripe seed, with a portion of its side cut out to shew the radicle at *a*.—7. A ripe fruit.—8. A nut; *a* the apex; *b* the base.

LETTER XXXVI.

THE HOUSELEEK TRIBE—PURIFICATION OF THE AIR BY PLANTS—THE SAXIFRAGE TRIBE.

Plate XXXVII.

Houseleek (Sempervivum tectorum) is a very common plant upon the roofs of cottages, and on old walls in the country. Its fleshy, starry leaves, are cooling and juicy; and, hence, the peasantry employ them as an application upon burns, or in other cases where the skin is inflamed. It is one of those species which are capable of growing in the most dry and exposed situations, often attracting its food from the atmosphere much more than from the scanty source that its roots have access to. It is usually planted by being enclosed in a lump of moist clay, which is stuck upon the naked tiles of a cottage. In such a situation, the young plant first secures itself by putting forth a few roots into the clay, and then gives birth to a number of little starry clusters of leaves, which surround their parent, and overshadow the place where the roots are to continue to develope: in the first instance, protecting it from the glare of the sun, and afterwards forming, by their decay, a soft vegetable mould, into which other roots may penetrate.

They are enabled to effect this by the power which

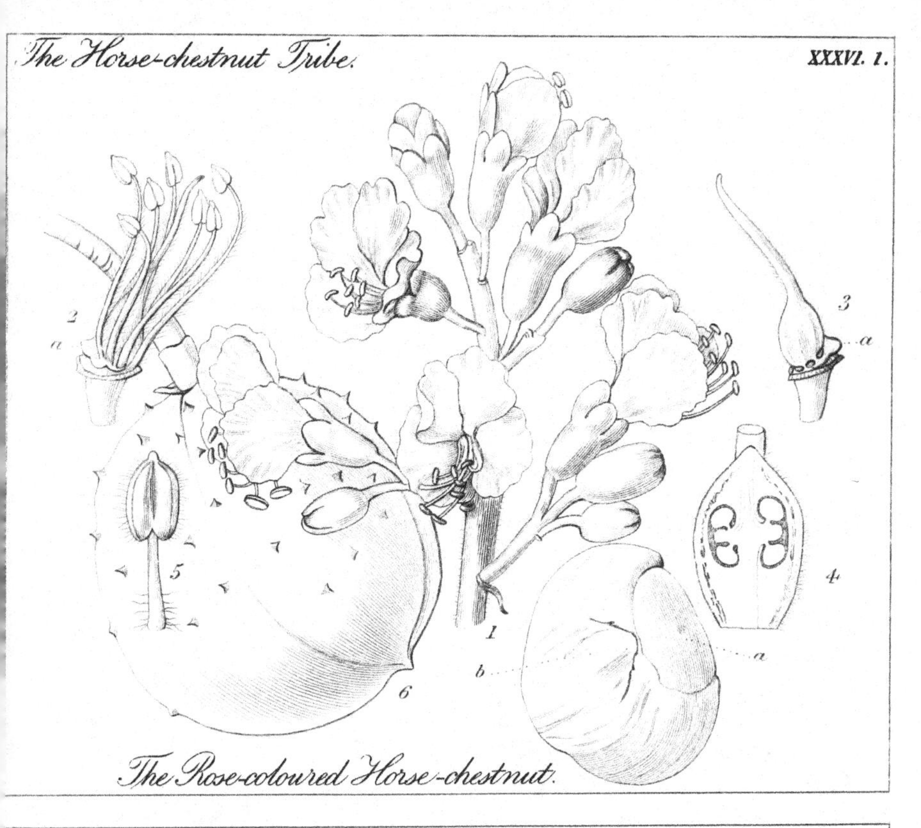
The Horse-chestnut Tribe.
XXXVI. 1.
2
a
3
a
5
4
1
a
b
6
The Rose-coloured Horse-chestnut.

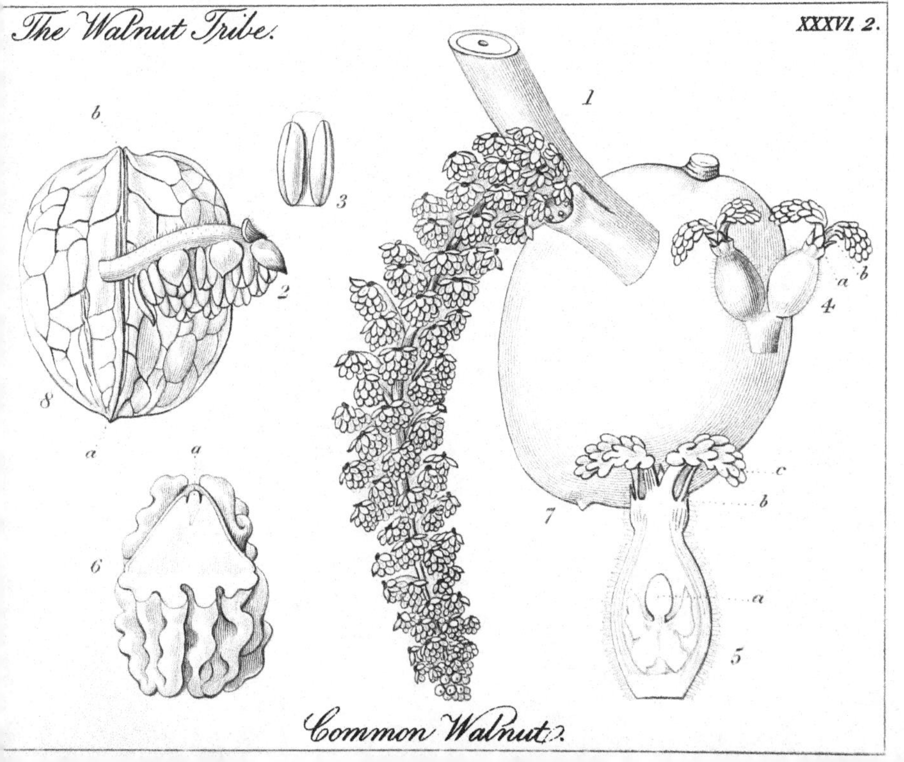
The Walnut Tribe.
XXXVI. 2.
b
1
3
2
a
b
4
8
a
a
c
b
7
6
a
5
Common Walnut.

they, in common with all other plants, but in a higher degree, possess of abstracting from the atmosphere its impure air, or carbonic acid, which they convert from a gaseous into a solid state, by separating the charcoal or solidifiable portion, and liberating the vital air or oxygen that was combined with it. By this wondrous process, living plants become the great purifiers of the air we breathe, and it appears quite certain, that if it were not for them the earth would soon become so pestiferous as to be uninhabitable.

All nature is in a continual state of decay and renovation. The perishing remains of animals and plants exhale putrid effluvia, which mix with the atmosphere and render it impure; the incessant action of respiration through the whole animal world, increases the impurity by abstracting the vital air or oxygen, and substituting foul air or carbonic acid. This combined action has been going on from the beginning of the present order of created things, and yet it does not appear that the air we breathe is less suited to our constitutions now than it was in the beginning. This is owing to the agency of plants, which, existing wherever animals or man can exist, are perpetually at hand to catch up and consume the impure particles of the atmosphere as fast as they are generated, and by fixing the carbonaceous part in their own systems, and again liberating the vital air or oxygen, with which the former was in combination, they restore to the air all the purity it had lost.

Here, then, you have another of those admirable proofs of wisdom and design that meet the philosophical observer at every step. Plants are Nature's eternal laboratories for the decomposition of all that would be injurious to man and other animals—the means by which the nicest equipoise is maintained between two most important opposite principles. Hence it is, that the most tiny blade of grass, or the most obscure weed, becomes in the hands of Providence an efficient means of working out the great design of the creation.

This is not a phenomenon liable to derangement or interruption, but arranged with the most admirable precision in every portion of its details. Thus, for example, although it is through the agency of leaves that the salubrious effect upon the air is brought about; yet we are not to suppose, that when the leaves have dropped from the trees, and the forest exhibits nothing but bare and naked branches, this agency is diminished. Leaves fall off indeed in winter, but at that time the corruption of the air, by the putrefaction of organized matter, is either arrested or very much diminished, and the green carpet which, even in the driest countries, springs up at that season, presents an elaborating surface of immeasurable extent, and amply sufficient to consume such gaseous impurities as may then be engendered. On the other hand, in the spring, when an elevated temperature sets rapidly at liberty the elastic impurities that the winter had bound in chains, leaves, too, are again produced with renewed vigour, and still carry off from the atmosphere all that the rapidly decaying matter is mingling with it, separating for themselves what man is incapable of respiring, and generating in its room in infinite abundance that vital air or oxygen, without which living things would perish.

> Hence, in bright floods, the Vital air expands,
> And with concentric spheres involves the lands;
> Pervades the swarming seas, and heaving earths,
> Where teeming nature breeds her myriad births;
> Fills the fine lungs of all that *breathe* or *bud*,
> Warms the new heart, and dyes the gushing blood;
> With life's first spark inspires the organic frame,
> And, as it wastes, renews the subtile flame.

These very beautiful lines are from the Botanic Garden of Darwin, a writer of an ingenious and philosophical turn of mind, whose poetry is now forgotten, although it has some splendid passages, and contains numerous descriptions of natural phenomena, expressed in language remarkable alike for its magnificence, and for its fidelity to what were, in the author's time, considered facts. Darwin, unfortunately, adjusted his natural phenomena to the unintelligible Rosicrucian machinery of gnomes, sylphs, nymphs, and salamanders, and this, together with the little knowledge that general readers possess of the facts his poetry was intended to illustrate, has been the cause of his poetical writings having fallen into neglect. I would, however, recommend you to read his Botanic Garden, especially the first part, called "The Economy of Vegetation;" you can easily pass by the tiresome Rosicrucian agency, and the remainder you will find extremely well worthy the perusal. But to return from our digression.

The property possessed by the common Houseleek, of growing on dry exposed roofs and walls, is participated in by a numerous kindred. In Teneriffe, where the genus Sempervivum is very common, the species, which are often shrubs of some size, not only occupy the steep cliffs and rocks in the neighbourhood of the sea, but actually, by their prodigious abundance, conceal the old gothic mansions of the interior of the island, overspreading the walls, and in the flowering season making them glow with the most brilliant golden tints; for the Houseleeks of Teneriffe have yellow flowers, while those of Europe have them of a rosy purple colour.

Such habits are indeed characteristic of all this tribe. In this country, the various races of Sedums, or Stonecrops, are constantly found in such situations; *Sedum acre* in particular, spreads its scaly stems and shining yellow starry flowers over the tops of walls in some places near London, and the *White Stonecrop* is equally abundant in others. An obscure little moss-like annual, Tillæa muscosa, overruns bleak, stony, naked commons, here and there; and on the grey stone walls of the valleys of the Wye and the Dee, and of the west and south-west of England, the graceful *Navelwort* (Umbilicus pendulinus) rears its delicate bells of green and gold.

Besides these plants, *Rose-wort* (Rhodiola rosea) puts up its purple heads of flowers in the woods, and by its terrestrial habit establishes the connection between the Houseleek tribe and the commoner forms of vegetation.

No tribe of plants can be more easily known than this; and the *White Stonecrop* (Sedum album, *Plate*

XXXVII. 1.) illustrates its structure perfectly. It has small, alternate, succulent, blunt leaves, between linear and oblong. Its flowers are white, and arranged in a compact cyme. The calyx (*fig.* 5.) is an olive-green, fleshy cup, delicately streaked with crimson, and divided into five, blunt, shallow lobes. The petals also are five, white, spreading, narrow, and sharp-pointed (*fig.* 2). Within these, from below the carpels, grow ten stamens, of which half are opposite the petals, and the other half opposite the lobes of the calyx. At the foot of each carpel (*fig.* 3.) there is a minute, yellow, flat, stalked gland, the end of which seems as if cut off (*fig.* 4.). The carpels have no adhesion to each other, are five in number, and stand opposite the petals, with which they agree in colour, size, and very much even in shape, except that they are rolled up, and taper much more to a point. The fruit is only a slight change from the flower ; the calyx and petals have lost their brillancy, are shrivelled and hang down (*fig.* 7.), the stamens are gone, and the carpels have assumed a pale brown hue ; they open at their inner edge (*fig.* 8.), and expose the seeds, which are small, smooth, and oval, and hang from their edges in a single row, upon short curved stalks (*fig.* 9.). The embryo (*fig.* 11.) is white, fleshy, and taper, and is tightly fitted by the seed-skin, its radicle pointing to the stalk of the seed.

The differences that mark the other British genera of the Stonecrop Tribe are easy enough to remember. *Tillæa* consists of minute moss-like plants, having only three or four petals and stamens, and no scales at the foot of the carpels. *Navel-wort* has the petals glued together by their edges into a little drooping bell. *Rose-wort* has only four petals and eight stamens. *Houseleek* has from six to twenty sepals and petals, twice as many stamens, and its scales are usually lacerated at the edge.

A consideration of the last mentioned plants necessarily leads to that of the *Saxifrage Tribe*, of which so many species occur in northern and mountainous countries, occupying the tops of walls, the sides, and even summits of mountains, the depths of wooded dingles, the sides of trickling streams, and even the recesses of the wildest bogs. They are remarkable for the exquisite neatness of their flowers, which are occasionally yellow or purple, but more generally snowy white, their pureness of colour being sometimes increased rather than destroyed by minute spots of the most clear and delicate crimson.

London Pride (Robertsonia umbrosa), which, although a native of the Yorkshire and Irish mountains, is so patient of smoke and impure air as to have derived its name from that circumstance, is one of the commonest species in cultivation, occurring in cottage gardens as frequently as daisies and primroses. You will know it by its round crenelled leaves, which are collected into little green roses, from the centre of which rises a graceful, reddish, branching panicle, the ends of whose slender branches are tipped by the most delicate little star-like flowers of pink and white. Another species (Leiogyne granulata) is common on banks and in hedges in May, peeping up from among grass and weeds, with its snow-white flowers drooping at the end of a long stem, scantily clothed with kidney-shaped few-lobed leaves. A third, the *Three-fingered Saxifrage* (Saxifraga tridactylites) springs up from the crest of walls, one of the earliest harbingers of spring. Let us take the latter for examination.

Three-fingered Saxifrage (*Plate* XXXVII. 2.) is a small annual, not much above three inches high, of a dull reddish brown in its foliage, which, as well as the stems and calyxes, is covered all over with glandular hairs of the same colour. Its lower leaves are divided into three tolerably regular lobes, whence its name; but those near the top of the little stems are undivided. The stem is quite unbranched, except near the top, where it divides into two or three forks, each of which is terminated by a single white flower. The calyx (*fig.* 2.) is oblong, and divided at the edge into five ovate lobes. There are five blunt white petals, originating from the side of the calyx; and ten short stamens placed also upon the calyx in a row after the petals (*fig.* 3.). The anthers are roundish flat cases, on short stiff filaments (*fig.* 5.). The pistil consists of a two-celled oblong ovary, which grows to the side of the calyx, almost to its top (*fig.* 3.), and then divides into two distinct, though short, styles, whose stigmas are little oval fringed spaces; you will remark that these styles are not only quite distinct from each other, but do not even spring from the same point, as is most usually the case in other plants. Each cell of the ovary contains a large convex placenta, all over which are placed minute ovules (*fig.* 3 and 4.).

The fruit (*fig.* 6.) is a seed-vessel covered by the glandular calyx, and opening at the point with two spreading valves; to its centre in the inside adhere the seeds (*fig.* 7.), which are exceedingly numerous, oblong, studded with elevated points (*fig.* 8. 9.), and contain an erect dicotyledonous embryo, enclosed in fleshy albumen (*fig.* 10.).

Such is the structure of the Three-fingered Saxifrage, and very nearly such is that of the principal part of its tribe, with the following very remarkable exception. In the plant just examined, the ovary adhered to the calyx for nearly all its length; such a circumstance, if occurring in one genus of a natural group, usually exists in all the remainder. But the Saxifrage tribe offers an exception to this rule; for in Leiogyne the seed-vessel is altogether free from the calyx, and in other cases it is partly free and partly adherent in the same genus.

This occurs in the genus Parnassia, one of the most curious of all wild plants, the companion of Sun-dew in her marshy haunts, and quite her rival in beauty and singularity of structure. The remarkable glands of Drosera are confined to her irritable leaves, and disappear in her flowers. In Parnassia, on the contrary, the leaves and stems are hairless, but there is a most extraordinary glandular apparatus in the flowers. The leaves of this plant are heart-shaped, and cluster round the base of the stem. The latter rises to the height of a few inches, bearing below its middle a solitary stalkless leaf, similar in form to those of the base, and on its point a single nodding white flower, whose petals are so beautifully marked by diverging sunken veins of a greenish colour, that a fanciful person might liken them to rivulets of chrysoprase flowing over a bed of snow. The glandular apparatus I have spoken of, consists of five fleshy scales, alternating with the stamens, and divided at their edge into numerous rays, each tipped with one beautiful pellucid greenish gland; so that the whole interior of the flower, when inspected from above, seems to bristle with a guard of fairy lances, tipped with sparkling jewels. I know of no natural object more exquisitely beautiful than this little flower, which you may cultivate for a few months by keeping its roots in wet bog-moss, and covering it with a bell-glass fully exposed to the light.

If you consider, as I hope you do, the resemblances of the tribes that are successively brought to your notice, with those which have been previously illustrated, you will have already noticed the near resemblance that exists between the Saxifrage and Rose Tribes. Not, indeed, between the Rose and the little plant we have just been looking at, but between it and the many herbaceous species that belong to the same group with the Rose. One of our usual contrasts will make this quite clear, and we may as well include in the comparison the Houseleek Tribe, which participates in the relationship of the Saxifrages.

I will first contrast their resemblances, and then their differences, in the same table, so that at one view you may perceive why they are placed near each other in the system, and why they are separated.

Saxifrage Tribe.	Houseleek Tribe.	Rose Tribe.
1. Leaves alternate.	1. Leaves alternate.	1. Leaves alternate.
2. Petals distinct.	2. Petals distinct.	2. Petals distinct.
3. Stamens growing from the side of the calyx.	3. Stamens growing from the side of the calyx.	3. Stamens growing from the side of the calyx.
4. Carpels more or less distinct.	4. Carpels more or less distinct.	4. Carpels more or less distinct.
5. Embryo as long as the seed.	5. Embryo as long as the seed.	5. Embryo as long as the seed.
1. Leaves sometimes with stipules.	1. Leaves without stipules.	1. Leaves usually with stipules.
2. Petals sometimes wanting.	2. Petals always present.	2. Petals sometimes wanting.
3 Carpels inferior or superior.	3. Carpels superior.	3. Carpels inferior or superior.
4. Carpels, when ripe, diverging and opening at the point only.	4. Carpels, when ripe, opening along their whole inner edge.	4. Carpels, when ripe, opening along their whole inner or outer edge.
5. Embryo in albumen.	5. No albumen.	5. No albumen.

Hence, it appears, that when the differences between these three tribes are strictly inquired into, there is nothing that will positively distinguish the Saxifrages from the Roses, except the albumen of the former, and the peculiar manner in which the two carpels spread away from each other, and open at the point when ripe.

As for the Houseleek Tribe, the distinctions by which it is known are more numerous and obvious, as you will see by studying the table.

EXPLANATION OF PLATE XXXVII.

I. The Houseleek Tribe.—1. A twig of *White Stonecrop* (Sedum album) in flower.—2. A flower magnified.—3. A view of the carpels and the scales at their base, the remainder of the flower being cut away.—4. One of the scales very highly magnified.—5. A calyx-cup.—6. A carpel, more magnified.—7. A ripe fruit, surrounded by the withered remains of the calyx and petals.—8. A view of a portion of the inner edge of a ripe carpel, shewing the manner in which the seeds are attached to its edges.—9. A seed.—10. The same cut across to shew the cotyledons.—11. The embryo.

II. The Saxifrage Tribe.—1. A tuft of *Three-fingered Saxifrage* (Saxifraga tridactylites).—2. A flower magnified.—3. The same divided longitudinally, shewing the situation of the stamens, and the interior of the ovary with its two styles.—4. A transverse section of the ovary.—5. A stamen.—6. A ripe seed-vessel.—7. The same divided longitudinally, to shew the placenta, to which a few seeds are still seen hanging.—8. 9. Seeds.—10. A seed divided lengthwise, with the embryo lying in the midst of albumen.

LETTER XXXVII.

THE BUCK-THORN TRIBE—SPINES—THE SPURGE TRIBE.

Plate XXXVIII.

YOU will sometimes see in curious gardens, you may always buy in the nurseries, or should you ever visit Greece or Palestine, you will find abundantly in wild rocky places, a spiny shrub, of a light and elegant aspect when it puts forth its new leaves in the spring, but of a savage withered appearance in the autumn, when its leaves are dried and discoloured, and its branches covered with a profusion of little, round, brown, flat seed-vessels, resembling ancient bucklers. This plant is called by the modern Greeks, Paliouri; by Botanists, Paliurus australis, or aculeatus; and by the English, *Christ's Thorn*, because it is said to have furnished the crown of thorns for our Saviour.

As this is a very interesting plant, we will take it for an illustration of the BUCK-THORN TRIBE, rather than the wild hedge-shrub, from which the latter derives its name. Its leaves (*Plate* XXXVIII. 1. *fig.* 1.) are alternately inserted upon slender, flexible branches; they are of an oblong figure, are slightly crenelled at the edge, and have three strong veins, which run from the one end to the other of the leaf, giving it a three-ribbed appearance. The leaves are

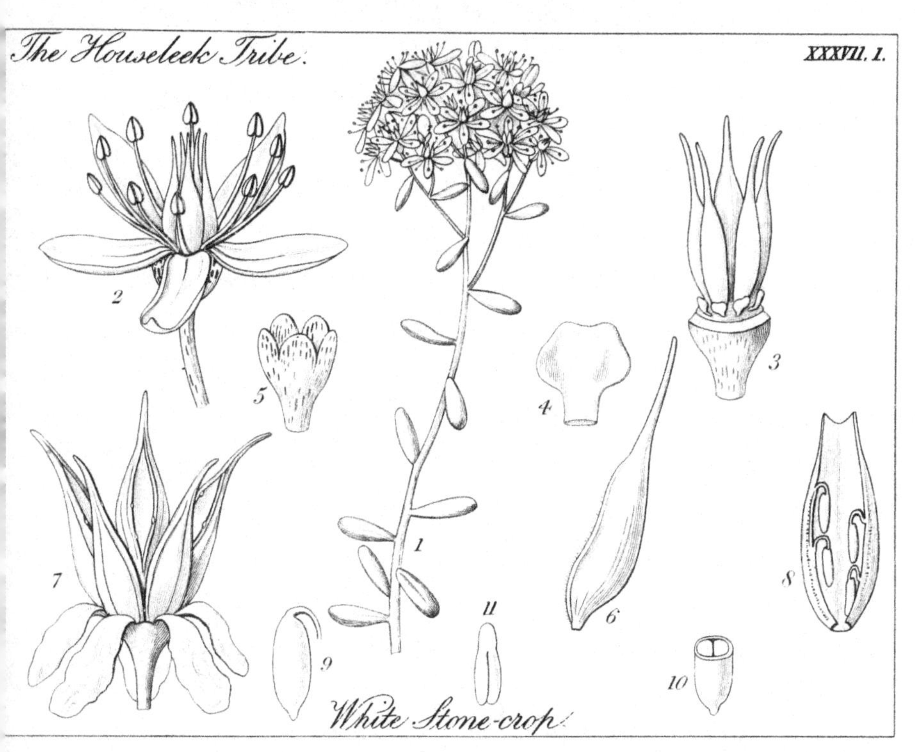
The Houseleek Tribe.
XXXVII. 1.
2
5
4
3
7
1
6
8
11
9
10
White Stone-crop.

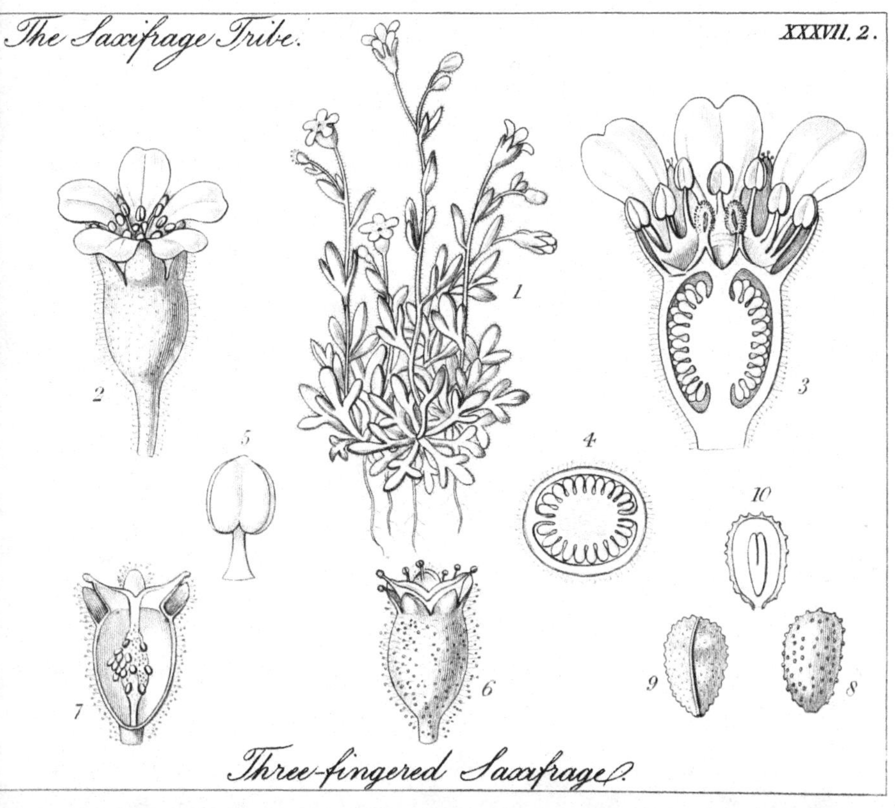
The Saxifrage Tribe.
XXXVII. 2.
1
2
3
5
4
10
7
6
9
8
Three-fingered Saxifrage.

placed on slender stalks, and in room of the stipules, which are characteristic of the greater part of the Tribe, they have a pair of sharp slender spines, which, upon the old branches, are curved outwards, and become so strong as to render hedges, made from the plant, perfectly impenetrable.

The flowers are pale greenish yellow, small, and grow in little stalked clusters, which are much shorter than the leaves themselves. They have a flat spreading calyx, divided into five sharp lobes (*fig.* 2. *b. b.*), each of which is a little raised in the middle, and accurately fitted to the edge of its neighbour before the flowers expand; so that, although the edges do not overlap each other at all, they nevertheless form a complete closed cavity, in which the stamens and petals are enclosed. This is one of the marks of the Buck-thorn Tribe, and is called a *valvate æstivation* (See Vol. I. *p.* 168.).

Alternately with the divisions of the calyx are placed five little yellow hoods (*fig.* 2. *a. a.*), which completely cover over the stamens; they are the petals. Opposite the petals are the five stamens; this is the next circumstance that you are particularly to attend to; there are very few instances where the stamens are opposite the petals, and also of the same number, and when it occurs it is always esteemed a structure of importance. After the stamens comes a broad, orange-coloured, lobed, flat disk (*fig.* 5. *a.* & *fig.* 2.), which does not touch the ovary, but simply lies upon the spreading sides of the calyx. The ovary is a little buried in the calyx at the bottom (*fig.* 5. *b.*), but otherwise is free; it contains three cells, in each of which is one ascending ovule; has a three-lobed style, and a glandular stigma at the end of each of the lobes.

This is all the preparation that is made for the singular fruit, which I have already described as resembling an ancient buckler, but which the French compare to a little head wearing a broad-brimmed hat, whence they call it *porte-chapeau*. To bring about the metamorphosis from the flower to the fruit, the following changes occur; the calyx—lobes, petals, and stamens drop off, and the branches of the style shrivel up; this reduces the flower to a roundish centre, surrounded by a flat-lobed limb. Then the disk, or limb, grows broader, the ovary swells, both change their appearance, the disk grows the fastest, the whole hardens and becomes brown, and the porte-chapeau (*fig.* 6.) is completed. It contains three cells, externally indicated by three low ridges, and in each cell there is a flat seed (*fig.* 7.). The seed contains an erect embryo, with two thin flat cotyledons, placed face to face, and a very short conical radicle (*fig.* 8.).

In considering the value of the characters thus described, as existing in the Christ's-thorn, you are to abstract—1. the valvate calyx; 2. the five stamens opposite the five hooded petals; 3. the fleshy disk; and 4. the three-celled, half-inferior fruit, with one upright seed in each cell; and you will have the characteristic features of the Buck-thorn Tribe. This is the more important for you to understand, because the Tribe comprehends species differing materially, in some respects, from what is found in the Christ's-thorn itself. For instance, few of the genera have a dry seed-vessel, but they more generally bear a succulent fruit; spines also are most frequently absent, or at least are alterations of buds, and not of stipules; and the leaves are most commonly not ribbed; but they all agree in the four characters just selected.

The *Alaternus*, one of those beautiful evergreen shrubs, which give such a peculiar charm to English garden scenery, *Buck-thorn*, so useful as a covert for game, and the *berry-bearing*, or *black Alder* of our copses, are various species of the genus Rhamnus, which is known from Paliurus by the fruit being succulent, the leaves ribless, and the stipules spineless. It contains several species of some importance for their dyeing properties; sap-green, for instance, is a preparation of the fruit of Buck-thorn (Rhamnus catharticus); the "French berries" of the shops, from which so beautiful a yellow is obtained, are the unripe fruit of the same plant; and yellow morocco leather acquires its colour from the juice of Rhamnus infectorius, and other southern species. The berries of all are unfit for food, and produce extremely unpleasant consequences when taken into the stomach.

Far otherwise is the case with the fruit of the *Jujube* (Zizyphus Jujuba), which, as I fear you know only too well, is mixed with some powerful gluten, and manufactured into lozenges, which are taken in coughs and colds.

Besides these, we have among the ornamental plants of the Jujube Tribe, the superb Ceanothus azureus, whose innumerable clusters of light-blue flowers have given quite a new character to our gardens in summer and autumn.

The spines of Christ's-thorn, remind me that I have never yet explained to you what spines really are. What they appear to be, I need not tell you; what they are, you may easily learn from a bush of the Sloe, on which they are sufficiently numerous. If you examine them, you will not fail to see that while a part are merely sharp hard points, others have a few buds upon their sides, and many more are invested with leaves, or even flower-cymes. They are, therefore, mere branches, with their points hardened and sharpened. Upon the use of spines, I find the following remarks by the late Professor Burnett:—"In barren, uncultivated tracts of heath, or common land, thorny plants abound, e. g. the Sloe (Prunus spinosa), the Rest-harrow (Ononis spinosa), the Hawthorn (Cratægus oxyacantha), the Buck-thorn (Rhamnus), the Cockspur-thorn (Cratægus crus Galli), and many others. These vegetables, when removed into gardens, and cultivated with care, lose all their thorns, which so thickly beset them when wild, and bear fruitful branches in their stead; becoming, as Linnæus expressed it, tamed plants (Plantæ domitæ), instead of the (Milites or) warriors, to use his language, that they were before. Willdenow was the first who explained the rationale of this metamorphosis, the first who shewed that thorns are abortive buds; buds which a deficiency of nourishment prevented becoming developed into branches, and which, when the requisite supply of food is present, speedily evolve their latent leaves and flowers. But Willdenow did not perceive the beautiful adaptation of means to ends, which forms, in my opinion, by far the most interesting part of the phenomenon.

"In open barren tracts of country, the very circumstance of the sterility of the soil must prevent the production of many plants, and of those which grow, few will be enabled to perfect many seeds. It is necessary, therefore, to protect such as are produced from extermination, by the browzing of cattle, otherwise not only would the progeny be cancelled, but also the present generation be cut off. And what more beautiful and simple expedient could have been devised, than ordaining that the very barrenness of the soil, which precludes the abundant generation by seed, should at the very same time, and by the very same means, render the abortive buds (abortive for the production of fruit) a defensive armour to protect the individual plant, and to guard the scantier crop which the half-starved stem can bear?"

These opinions are borrowed from Darwin (*Botanic Garden*, Vol. ii. 139), and are ingenious enough. I am, however, by no means sure that they are well founded. But with objections to them, I am not disposed to entertain you.

Of course you will not confound the spines or thorns of the Buck-thorn, the Christ's-thorn, the White-thorn, the Black-thorn, &c. with the prickles of the Rose, because the latter are also popularly called thorns. True spines or thorns grow from the wood of plants; prickles, or false thorns (*aculei*), grow, like hairs, from the surface of the bark.

Another Tribe, related to the foregoing, is that of the Spurges (*Plate* XXXVIII. 2.); plants distinguished from all others by two characters; the one that of having the stamens in one kind of flower, and the pistil in another, the other that of having a fruit which divides, when ripe, into three coccoons, whence it is called tricoccous. By these peculiarities are combined a large number of exceedingly remarkable plants, many of which are highly deleterious, most of which are exotics, and a very small number of which are either wild in our woods, or cultivated in our gardens.

Among them, few are more remarkable then the *Palma Christi* (Ricinus), with its deeply-lobed, livid, purple leaves, and long clusters of stamen-bearing flowers, at the base of which are clustered a few spiny pistilliferous ones. Another species is the *Box-tree; Tapioca* and *Cassava* are yielded by a third (Jatropha Manihot); and *Indian rubber*, that curious substance, to whose utility there really seems to be no limit, flows from the wounded bark of others. Arrows are poisoned with the dangerous juice of various species; and there is a long succession of them upon the list of fatal or useful plants.

Few plants are more remarkable for their properties, than *Manchineel* (Hippomane Mancinella).

> "If rests the traveller his weary head,
> Grim Mancinella haunts the mossy bed,
> Brews her black hebenon, and stealing near,
> Pours the curst venom in his tortured ear."

It is a West Indian tree, with which the Indians poison their arrows; and the dew that falls from it is reputed to be so caustic as to blister the skin, and produce dangerous ulcers; whence many persons have found their death by only sleeping beneath its branches.

This statement is contradicted by some writers, and doubted by others; but there is no sufficient reason for calling it in question. It is perfectly certain that the juice, when applied to the skin, produces a pain like that of red-hot iron, as is proved by the infamous practice of slave-drivers having steeped their scourges in Manchineel juice, before they flogged their negroes.

We have no wild plant that well illustrates the structure of this order, except the common Box. But we have a most common genus, that to a certain degree explains it, and which has a singular structure of its own. This, therefore, which is the common Spurge or Euphorbia, I have selected for illustration.

The common *dwarf Spurge* (Euphorbia Peplus, *Plate* XXXVIII. 2.) is an annual, with a slender, smooth, branching stem, which discharges in profusion a milky juice when wounded. It is a general property of its tribe to do the same. Its leaves are obovate, tapering to the base, stalkless, and placed in a ring of three, immediately below the branches that bear the flowers. The leaves of the flower-branches are differently shaped from those of the stem, opposite in pairs, ovate with a heart-shaped base, and sharp-pointed.

The flowers either grow in the forks of the branches (*fig.* 1. *a. a.*), or among the uppermost leaves singly. They are green cups (*fig.* 2.), of a most curious conformation. The edge of the cup is divided into ten lobes, of which five are flat, spreading, glandular, and two-horned (*fig.* 2. *a. a.* and *fig.* 3. *a. a.*), and five scale-like, inflected, and fringed with hairs (*fig.* 3. *b. b.*). From the very bottom of the cup rises a cluster of stamens, of unequal lengths, each having a joint in the middle (*fig.* 4. *a.*); these stamens rise up one by one, or in very small numbers, protrude themselves beyond the mouth of the cup, to discharge their pollen, and then shrivel up. From their centre springs a long, green stalk (*fig.* 2. *b.* and *fig.* 3. *c.*), curved downwards by the weight of a roundish ovary that grows upon its summit. There is a joint in the stalk of the ovary of the same nature as that in the stamens.

The ovary (*fig.* 6. & 5.), is three-cornered, has a double short wing at each angle, and contains one pendulous ovule in each cell; two stigmas, or rather a two-lobed stigma, rises from each lobe of the ovary. The seed-vessel is of the same form as the ovary, and separates with elasticity, when ripe, into three cases, or cocci, out of each of which falls a single seed. The seeds are slightly downy, pale straw-coloured, faintly spotted with purple, and unequally six-sided (*fig.* 5.); next the hilum, they have a white fleshy protuberance, called a *caruncula*, and they contain an embryo with two short cotyledons, and a long slender radicle lying in fleshy albumen.

What now is the real nature of the parts we have been examining? It used to be thought that the green cup was a calyx, and that the stamens were of the same nature, exactly, as other stamens. But it was remarked, in course of time, when more exact views of Botany began to be entertained, that a joint in the apparent filament was seen nowhere else, that another in the stalk of the ovary was equally unusual; that from this joint there sometimes springs a sort of cup-like membrane; that the confused arrangement of the stamens was extremely unlike the regularity with which those parts are usually inserted; and, finally, that no other genus could be found in the tribe of Spurges, in which the stamens and the pistil occur in the same flower. These considerations led to the discovery that the cup is an involucre, with a glandular and lobed border, that each stamen is a single flower, consisting of a single stamen, without either calyx or corolla, the place of those organs being indicated by the joint in their middle, and that the ovary in the centre is, in like manner, a single, separate flower; so that the apparent flower of a Spurge is in reality a curious kind of flower-head.

Thus you see, that even in so humble and insignificant a weed as this, there is much to study and admire. In general, the species of Euphorbia are possessed of but little beauty, but there are some remarkable exceptions; for their floral leaves, and their cups, or the glands upon them, become in certain cases coloured of the most vivid tints, scarlet, crimson, emerald-green, or white, and as the parts are usually enlarged in proportion, a most brilliant effect is occasionally produced, notwithstanding the universal want of calyx and corolla in this tribe.

EXPLANATION OF PLATE XXXVIII.

I. THE BUCK-THORN TRIBE.—1. A twig of *Christ's-thorn* (Paliurus australis).—2. A flower magnified; *a a* petals, *b b* lobes of the calyx.—3. A petal separate.—4. A stamen.—5. A section of the ovary and disk; *a* the disk, *b* the part where the ovary grows to the side of the calyx.—6. A ripe fruit, natural size.—7. A seed.—8. A section of the same, exhibiting the flat embryo.

II. THE SPURGE TRIBE.—1. The upper part of the stem of *Dwarf Spurge* (Euphorbia Peplus), with the common leaves at the bottom, and the floral leaves occupying the remainder of the specimen; *a a* flower-heads.—2. A flower-head, magnified; *a a* glandular divisions of the involucre; *b* the pistil-bearing floret in the centre.—3. A section of the involucre; *a a* the glandular lobes; *b b* the inflected lobes; *c* the stalk of the pistil-bearing central floret, surrounded by the naked stamen-bearing florets.—4. A stamen-bearing floret; *a* the joint between *b* the pedicel, and *c* the filament.—5. A section across a nearly ripe fruit, shewing the short wings at the angles, and the seeds with the embryo lying in the centre of the albumen of each.—6. The same ovary, uncut.—7. A seed; *a* its caruncula.—8. A longitudinal section of the same, with the embryo surrounded by albumen.

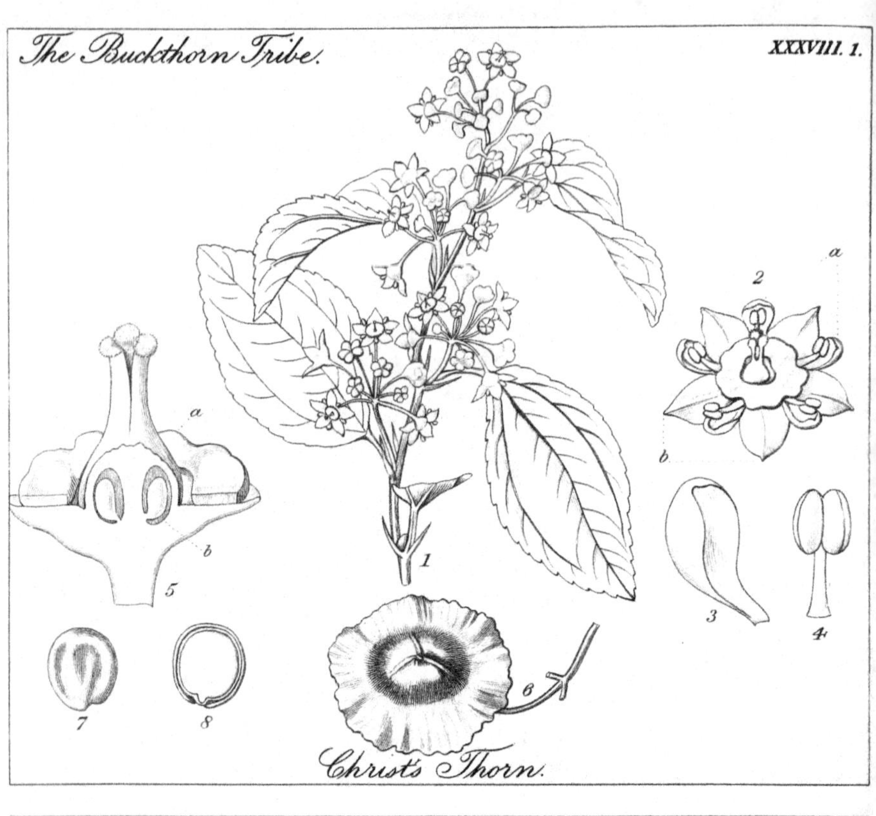
The Buckthorn Tribe.
XXXVIII. 1.
a
2
b
a
b
5
1
3
4
7
8
6
Christ's Thorn.

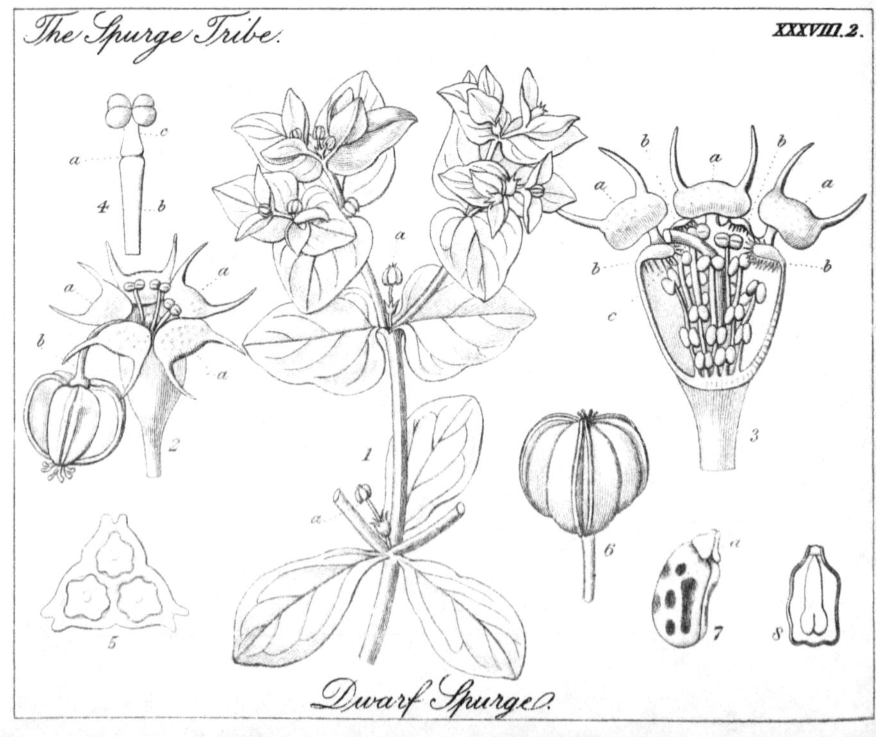
The Spurge Tribe.
XXXVIII. 2.
c
a
4
b
a
a
b
a
2
b
b
a
b
a
a
b
b
c
3
1
a
6
5
7
8
Dwarf Spurge.

LETTER XXXVIII.

THE FLAX TRIBE—ABORTIONS—LINEN
THE RUE TRIBE.

Plate XXXIX.

Among the plants that are grown in fields, for their utility, the prettiest, I think, is Flax, with its nodding blossoms, and its light-blue petals, which, day after day, during the flowering season, continue to strew the soil with azure fragments. It was once considered a member of the Chickweed Tribe, but if you compare it with the species of that group, you will wonder, not that it is now separated, but that it should ever have been associated with them.

In the first place, its stems and leaves are quite different; the joints of the former are not swollen, and the latter are not opposite (*Plate* XXXIX. 1. *fig.* 1.). Secondly, its calyx has the sepals in a broken whorl (*fig.* 2), two external, and three internal, which is not at all the character of the Chickweeds; moreover, the ovary contains ten cells, in each of which is one pendulous ovule (*fig.* 5.); and finally the seed-vessel splits into ten sharp-pointed valves (*fig.* 6.). These circumstances are considered sufficient to elevate the Flax into the type of a natural assemblage, consisting of scarcely any other genus; and accordingly the Flax

Tribe is now admitted into the works of all systematic writers.

Its principal points of agreement with the Chickweed Tribe consist in its having five petals, five stamens growing below the ovary (*fig.* 3.), and five distinct styles; all points of slender importance in themselves, and in the present instance quite neutralized by the nature of the predominating differences above explained.

It might have been more correctly allied to the Mallow Tribe, for you will remark that its stamens grow into a tube (*fig.* 3.), that it has pin-headed, or, as we say, capitate, stigmas, and several one-celled carpels, arranged in one whorl round an imaginary axis; it moreover agrees with that group in possessing mucilaginous properties. But, on the other hand, its leaves have no stipules, its calyx is extremely different from the valvate one of Mallows, and it has not their crumpled folded embryo.

In fine, it is rather to the Rock Rose Tribe, and the plants assembled in that vicinity, that Flax must be compared, as you will hereafter see.

Among the peculiarities of Flax, that do not belong to its character as a distinct natural group, but that are exclusive to the genus Linum, of which it is a species, are two that deserve particular notice; the one, the abortion of half its stamens, the other, the occurrence of a ten-celled ovary, in connection with five styles.

You will remark that the five stamens of Flax are united by their base into a downy cup (*fig.* 3.), and that five small teeth (*fig.* 3. *a. a.*) alternate with them.

The teeth are the rudiments of stamens, and show that there is a tendency in Flax to produce ten stamens, but that, owing to some unknown constant cause, only five of them are actually developed. This disposition to form parts, without actually forming them, is what Botanists call *abortion;* and is one of the most common of all phenomena. The knowledge of the fact is of great importance, because it helps us to reconcile apparently contradictory circumstances, and to reduce, within fixed rules, the laws that regulate the innumerable modifications and combinations of the organs of plants. One or two examples will make this clearer to you.

It is an established axiom that the divisions or parts of each successive whorl of organs, are placed alternately with those which succeed them. Thus the following arrangement of letters will show the successive positions of the parts of a flower that consists of five sepals (S), five petals (P), five stamens (s), and five carpels (c); provided the parts were placed in parallel rows instead of concentrically—

<table>
<tr><td>S</td><td></td><td>S</td><td></td><td>S</td><td></td><td>S</td><td></td><td>S</td><td></td></tr>
<tr><td></td><td>P</td><td></td><td>P</td><td></td><td>P</td><td></td><td>P</td><td></td><td>P</td></tr>
<tr><td>s</td><td></td><td>s</td><td></td><td>s</td><td></td><td>s</td><td></td><td>s</td><td></td></tr>
<tr><td></td><td>c</td><td></td><td>c</td><td></td><td>c</td><td></td><td>c</td><td></td><td>c</td></tr>
</table>

so that the stamens would be opposite the sepals, and the carpels opposite the petals.

But if the number of petals were ten instead of five, the position of the stamens, with respect to the sepals, would be altered, and the latter would be opposite the first or outer row of petals, thus—

```
S       S       S       S       S
    P       P       P       P       P
P       P       P       P       P
    s       s       s       s       s
c       c       c       c       c
```

And other changes in proportional numbers would be productive of corresponding alterations of position. These differences are found of great importance in systematic Botany, and every good writer pays the most careful attention to them. Their value is very much increased by considering the nature and degree in which abortion takes place, and observing, by the manner in which it affects the usual order of succession, whether it indicates a tendency to the production of more rows of parts than actually develope, or to the suppression of a portion of those rows that are in part completed, and what relation is really borne to each other by the parts that appear, and those that do not develope. If you look into this subject practically, you will find that the abortion of particular organs, or rows of organs, is, in a great number of cases, the most unerring sign by which certain natural groups are distinguished, and that the importance of the abortions, in a systematic view, is in proportion to the degree in which they derange the symmetry of the flower, or cause a deviation from regular structure. I cannot do better than give you several instances of this, by tables similar to the preceding, in which the letters have the same value as before; those which indicate partial abortions being printed in italics, and in a smaller type, and total abortions being represented by dots.

Let us begin with Flax itself, which offers but a slight instance of abortion. Its parts are thus—

S S S S S

P P P P P

s s s s s

s *s* *s* *s* *s*

c c c c c

Here the consequence of this presence of the abortive row *s*, is to throw the carpels out of their place, and to bring them opposite the sepals, instead of opposite the petals.

The Primrose Tribe shows a deviation of a more important nature. It calyx, corolla, and stamens are thus—

S S S S S

P P P P P

. . . . .

s s s s s

In this case the third row, whether belonging to the petals or stamens is missing, and the consequence is that although the stamens are equal in number to the petals they are opposite to them, instead of alternate. This also happens in the Buckthorn Tribe, and elsewhere, but I have chosen the Primroses to illustrate this sort of irregularity, because that tribe, in the instance of the genus Samolus, contains a proof that the abortion which theory points out really does exist. Its parts are thus—

S S S S S

P P P P P

s *s* *s* *s* *s*

s s s s s

These cases, however, are nothing to what occurs from abortion in many Endogens. The greater part of the Orchis Tribe is thus—

```
S         S         S
     P         P         P
.         s         .
```

The Arrow-Root Tribe thus—

```
S         S         S
     P         P         P
p         p         .
     .         .         s
```

The Ginger Tribe thus—

```
S         S         S
     P         P         P
                    p
.         .
     s         s         s
```

The Banana Tribe thus—

```
S         S         S
     P         P         P
s         s         s
     s       s or .      s
```

which is very nearly in accordance with the ordinary structure of Monocotyledonous groups.

Perhaps, however, there is no more curious case of extensive alteration in structure, in consequence of abortion, than in the Mint Tribe, of whose flowers the following letters express at once the theoretical and real composition—

```
S       S       S       S       S
    P       P       P       P       P
            or      or
            p       p
s       s       .       s       s
        or              or
        .               .
    .       .       .       .       .
.       .       c       .       c
```

In the Mint Tribe it is especially to be remarked that each carpel is divided into two lobes, so that, although there are four external partitions in the ovary, yet there are only two carpels, which, in fact, correspond with the two lobes of the style. That three other carpels are undeveloped, is proved by certain cases in which they are actually present, in addition to the two ordinary ones; in such instances the ovary consists of ten lobes, and the style is divided into five little segments.

This fact brings me back to the second subject, which, I have already said, deserves particular notice in the Flax; namely, the ten cells of the ovary, and the five styles. I need scarcely now repeat, that, under all circumstances, the number of styles corresponds with the number of carpels of which the pistil is composed, or of the lobes of the stigma when the styles are all consolidated, provided any lobes are discoverable. As in the Mint Tribe, under ordinary circumstances, there are four lobes of the ovary, and two lobes of the stigma, it therefore follows, that each carpel is two-lobed; and I have just explained that certain monstrous cases prove that such is really the fact. Now suppose that two such lobes are consolidated, we then have carpels each with two cells, as in the Vervain Tribe, and this is only what we find in the Flax. You will observe, however, that although in the latter plant there are two cells to each carpel, yet the dissepiment that divides them is imperfect (*fig.* 5. *a a*); so that, although, for the purpose of illustration, I have supposed that each carpel of the Flax may be formed by the consolidation of two lobes, yet it is more probable that in reality its peculiarity is simply owing to the projection of a short plate from the back into the cavity of each cell.

Common Flax (Linum usitatissimum), as its name imports, is the plant from which *linen* is manufactured. Its stems are soaked for a long while in water, until the cellular substance rots away, and then the tough fibres that remain behind are cleaned, dressed, and converted into linen thread. You are doubtless aware of the great superiority of linen over cotton thread, in regard to durability and toughness. This is owing to the different nature of the organized substance from which they are prepared. The part of the Flax that remains after maceration is its woody tubes, the toughest and strongest part of the vegetable fabric, and that to which all plants owe their flexibility and strength. It is the part which enables the leaf to bear the violence of the storm without injury, which gives its value to timber, and which enables the cane and the lancewood to bend so freely without breaking. Cotton, on the contrary, is merely the hair that grows upon the seed of the Cotton plant, and is a form of that cellular substance which constitutes the parenchyma of leaves, the delicacy of flowers, and the pulpiness of fruit, which fills up the interstices between the woody tubes, and holds together the sinewy framework of vegetation.

Garden Rue (Ruta graveolens, *Plate* XXXIX. 2.) is the type of a very extensive natural group, called, after it, the Rue Tribe. It consists of plants having a powerful, and usually a nauseous, odour, and their leaves filled with transparent dots (*fig.* 7.), in consequence of their secreting an essential oil, which renders them valuable in cases of spasms.

Rue itself, *Fraxinella*, covered with fragrant glands, which are said to exhale their volatile parts in such abundance in hot weather as to render the atmosphere that surrounds it inflammable, and different sorts of *Diosma* and *Correa* are those which are most common in gardens. The remainder are principally exotics, which are little known in cultivation. Rue itself will give you a good idea of their general nature.

It is a perennial, hairless, glaucous plant, having a strong, peculiar, disagreeable odour. Its leaves are unequally pinnated, rather fleshy, crenelled, and dotted like those of an Orange. The flowers are greenish-yellow, and grow in cymes at the end of the branches. The calyx (*Plate* XXXIX. 2. *fig.* 2.) consists of four spreading, toothletted sepals. There are four petals, with short claws, and a very concave toothletted end. Eight spreading stamens arise from a fleshy ring surrounding the ovary, and having about sixteen pits impressed upon it, in a circle, a little above the origin of the stamens (*fig.* 2.). Upon this ring is planted a conical, four-lobed, uneven ovary, consisting of four cells, which are not parallel, as usually is the case, but spread away from each other at the base, around a fleshy elevated centre (*fig.* 3. *a.*). Altogether the mass of fleshy matter, upon which the cells of the ovary are placed, is so considerable as to have in systematic Botany a particular name, that of *gynobase.*

The cells of the ovary contain about four ovules, placed upon a prominent placenta (*fig.* 3. *b.*). The style rises from between the points of the lobes of the ovary, and is divided at its apex into four obscure teeth. The seed-vessel is a light brown dry capsule, splitting into four coccoons (*fig.* 4. *a.*), in each of which is a single seed, and which surround the thickened hardened gynobase (*fig.* 4. *b.*). The seeds are dark brown, pitted, angular bodies (*fig.* 5.), containing an embryo lying in the midst of fleshy albumen (*fig.* 6.).

Such is the structure of Garden Rue, and the same is found more or less in the numerous genera referred to its tribe. As you are little likely to meet with many of them, I will only remark that some are curious, as Correa, for having their petals united into a tube, like that of a Monopetalous plant, and that they do not differ much from the Orange Tribe (Vol. I. *p.* 86. *Plate* VI. 2.), except in their dry splitting fruit, their great fleshy gynobase, and their albuminous seeds.

EXPLANATION OF PLATE XXXIX.

I. The Flax Tribe.—1. A twig of *perennial Flax* (Linum perenne).—2. A magnified flower, from which the petals have dropped off.—3. The stamens and styles ; *a a* teeth representing abortive stamens.—4. The ovary with the bases of the five styles.—5. A section of the ovary; *a a* the imperfect dissepiments.—6. A ripe seed-vessel, with its valves separated.—7. A seed.—8. A section of the same, shewing the embryo.

II. The Rue Tribe.—1. *Garden Rue* (Ruta graveolens).—2. A magnified flower, without the petals; *a* the ring of pits above the stamens.—3. A section of the ovary; *a* the gynobase, *b* the placentæ. —4. A seed-vessel, from which the seeds have fallen; *a* the coccoons, *b* the gynobase.—5. A ripe seed.—6. The same cut longitudinally to shew the embryo and albumen; *a* the hilum.—7. The tip of a leaf, with its pellucid dots.

LETTER XXXIX.

THE BUCKWHEAT TRIBE—THE GOOSEFOOT TRIBE.

Plate XL.

No plants are more common by road-sides, and in waste places, than the species of the genera Polygonum and Rumex; or in flower-gardens, than Persicarias; or in the kitchen garden, than Sorrels and Rhubarbs. These plants belong to the BUCKWHEAT TRIBE (Polygonaceæ), and will next demand our attention.

You will, no doubt, remember the Nettle Tribe (Vol. I. *Plate* XI. *p.* 147.), with its hairy calyxes without petals, its flowers of two sorts, and its single ovary, containing one upright seed; nor do I imagine the gorgeous Amaranths to be forgotten, whose calyx is so much like that of the Nettles, only that its colours are so gay, its flowers all of the same sort, and the leaves without stipules. The Buckwheat Tribe is in many respects like these, but at the same time essentially different. Take for an example *Knot-grass* (Polygonum aviculare), species of which are sure to be met with on every neglected garden walk, or hard bank, where few other plants could exist at all. There it expands its numerous slender arms, embracing the hard earth, and pressing to its bosom the cold rock

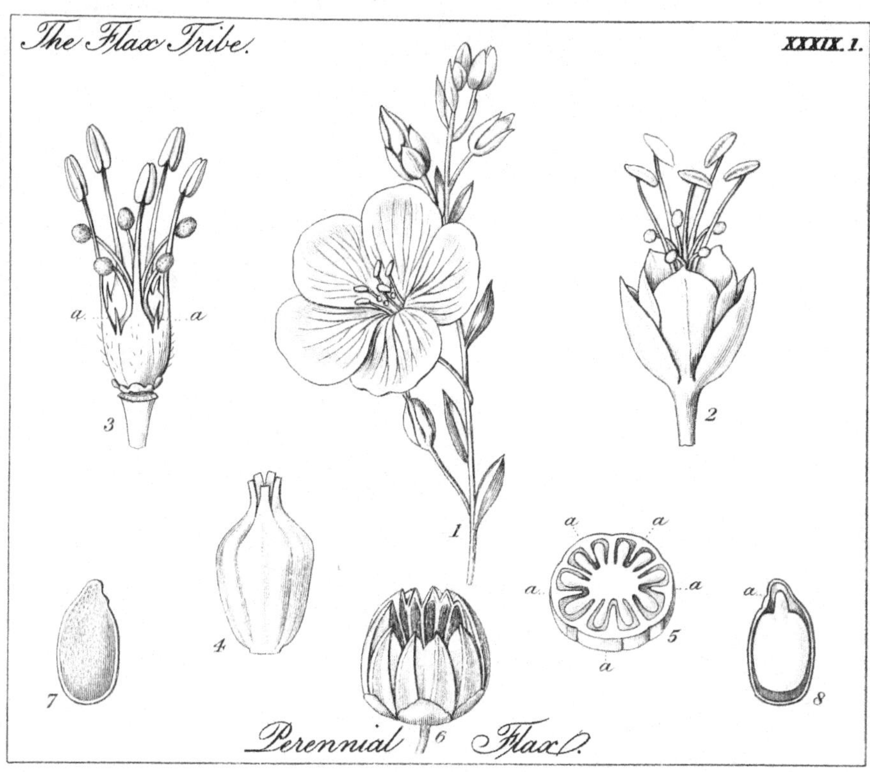
The Flax Tribe.
XXXIX. 1.
a
a
3
2
1
a
a
a
a
a
a
4
5
7
8
6
Perennial Flax.

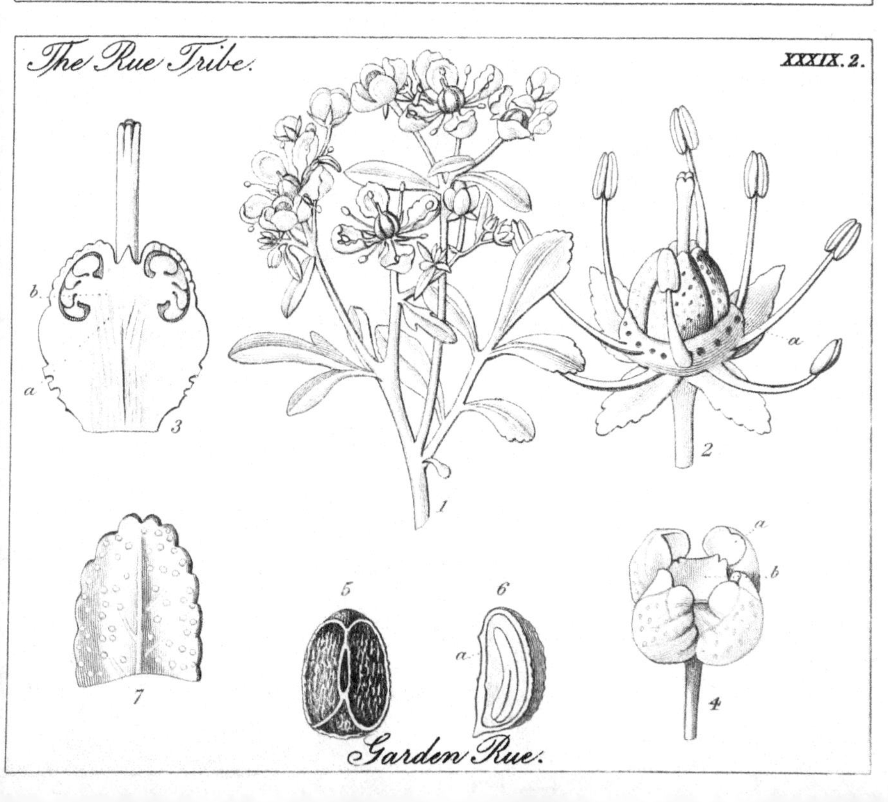
The Rue Tribe.
XXXIX. 2.
b
a
3
a
2
1
a
b
5
6
a
7
4
Garden Rue.

on which nothing else can grow, equally regardless of hunger and parching thirst. Nay, do not start at this strange description; it is literally as well as figuratively true.

Knot-grass (Polygonum aviculare, *Plate* XL. 1.) is one of the commonest of European weeds; whereever a seed can take root in the neighbourhood of man, and where nothing else, not even a Stonecrop, can fix itself, there you will find Knot-grass, lying prostrate on the soil, and continually spreading away from a common centre. Its stems are slender, and wiry; its leaves are narrow and oblong, with a curious pair of fringed, ribbed stipules at the base (*fig.* 2. *a.*), surrounding the stem, and forming a sort of tube through which its joints pass; such stipules, which are very uncommon, have obtained the technical name of *ochreæ* or boots. The flowers are sessile and axillary in the bosom of the leaves. They consist (*fig.* 3.) of a calyx divided into five imbricated parts, which unite at the base in an herbaceous tube. Into the throat of the tube are inserted seven stamens (*fig.* 4.), of equal length, but having no certain position with respect to the lobes of the calyx; constituting however, in theory, almost one whorl and a half. The ovary (*fig* 5.) is an oblong, three-cornered body, with three separate stigmas, and one erect ovule in its inside. The fruit (*fig.* 6.) is a three-cornered, hard, deep-brown nut, encircled by the calyx, and containing a curved embryo lying on one side of some mealy albumen (*fig.* 7.), the radicle of the embryo nearly touching the apex of the seed.

It therefore differs from the Nettles in having booted stipules, uniform flowers, and triangular fruit, and from the Amaranths, in having stipules, triangular fruit, and an inverted embryo. There is nothing else within your acquaintance with which it is necessary to compare it. Hence the Buckwheat Tribe, the species of which, however different from Knot-grass, agree with it essentially, is a peculiar natural order, cut off by strong lines of demarcation from all that surround it.

Knot-grass itself, I have already said, is a species of Polygonum, and there are many other wild plants belonging to the same genus; of these P. hydropiper and Persicaria, with their short, rounded spikes of pink calyxes, are common examples; in the gardens P orientale, or *Garden Persicaria,* with its crimson panicles, is one of the showiest of annuals; and in the fields *Buckwheat,* or *Beechwheat* (Polygonum Fagopyrum), so called from the resemblance of its little hard-brown seed-vessels to Beech-mast, with its beautiful rose-coloured flowers, is commonly cultivated for its seeds, of which pheasants are remarkably fond, and from which is prepared the flour from which in part crumpets are made.

But these are far from all; *Docks,* the detestation of the farmer, with all their hedge varieties or species, and *Sorrels,* which the French cooks value so much, notwithstanding their unwholesome acidity, are different species of Rumex; while Rheum boasts of the useful *Rhubarbs,* whose leaf-stalks afford a pleasant substitute for gooseberries in the early spring, and of the drug of that name, which is one of the greatest preservatives that nature has provided for the delicate machinery of man.

Many other exotic plants equally belong to this tribe; but they are not worth the introduction here.

Very closely allied to the plants last mentioned, are those which constitute the GOOSEFOOT TRIBE (*Plate* XL. 2.), a natural order comprehending such culinary vegetables as *Spinach* (Spinacia oleracea), *Orach* (Atriplex hortensis), *Beet* (Beta vulgaris), and the like. They are plants whose flowers are of an herbaceous or dull red colour, and of a succulent texture, so that they all are, without exception, unattractive species. What Nature has denied to their flowers, she sometimes however gives to their leaves, which are occasionally stained with the most vivid tints of yellow, purple, crimson, and even rosy red, as in the Chard Beet, and the Garden Orach. Most, and perhaps all, are suited for cooking as spinach, in consequence of the pulpy, tender, sub-mucilaginous quality of their leaves; but, as they differ in quality, Spinach itself is generally preferred for the table; none of them, however, are better than the wild *Sea Beet* (Beta maritima), which loves to fix itself on the sea-shore at the foot of chalky cliffs, often within reach of the spray.

One of the commonest species is *Goosefoot* (Chenopodium album), a grey, powdery, annual weed, which springs up on every heap of rubbish, and soon produces at the ends of its upright branches numerous clusters of minute green flowers (*Plate* XL. 2. *fig*. 1.), which, without changing colour, ripen their seeds, and drop them in profusion on the surrounding soil. It is scarcely possible to select a plant more unattractive than this, and yet, if you will attentively study its structure with me, you will find that, in some respects, its beauty is of no common order.

I have already said that it is an annual; its stems are angular, and grow about a foot and a half or two feet high, producing a few stiff upright branches. The leaves are of a dull grey green on the upper side, and of a dead glaucous colour on the under side; they are, moreover, powdered with a loose mealy substance, which spreads indeed over all the parts of the plant exposed to the air, and which seems to be a peculiar cutaneous secretion. Viewed under the microscope, and illuminated by a ray of bright light thrown from above, this secretion gives the plant the most beautiful glittering appearance, every part of the surface being spangled with what seem fragments of emeralds and chrysoprases. The leaves are placed on short stalks, and they have a somewhat lozenge-shaped figure, with several coarse toothings on their edge.

The flowers are arranged in compact clusters, proceeding from the axils of the leaves. Each, when unexpanded (*fig*. 2.), is round, depressed, marked with five prominent, rounded angles, bright green, exquisitely studded in the hollows between the angles with little glittering balls, which have all the appearance of being consolidated dew; indeed the whole flower-bud has much the aspect of a tiny green *sea-egg*

(Echinus). All this pretty show belongs to the calyx, which finally unfolds into five, spreading, pale green lobes *(fig.* 3.), with a white pearly border. There are no traces of petals. The stamens are five, slightly adhering by their bases into a very shallow cup, and placed opposite the divisions of the calyx. The ovary *(fig.* 4.) is roundish, superior, with two long hairy stigmas, it is one-celled, and contains a single ovule, attached to the bottom of the cell by an oblique cord *(fig.* 5.). The seed-vessel is a thin semi-transparent bag, which breaks irregularly when ripe, and drops a single jet-black flattish seed *(fig.* 6.), containing an embryo, curved round mealy albumen, and pointing its radicle to the hilum *(fig.* 7.).

In general, plants of the Goosefoot Tribe are so similar to this in structure, as to give the student no trouble in identifying them ; some, for example, have the stamens in one flower, and the pistil in another, as Spinage; others have the base of the calyx hardened and partially adhering to the ovary, as Beet ; but such differences are no greater than occur in all natural orders. There is, however, a wild plant belonging to the Goosefoot Tribe, which is so curious in its appearance, as to deserve particular mention. It is often brought, from the salt-marshes where it grows, to market, under the erroneous name of Samphire,* and being prepared with spice and vinegar, forms a coarse kind of pickle. At first sight you would take this plant, whose real

* The real Samphire is an Umbelliferous plant, found on the chalky cliffs of our southern coast ; it is the Crithmum maritimum of Botanists.

name is *Glass-wort* (Salicornia), to be leafless and flowerless, with nothing but jointed brittle stems; for its shoots really look as if they were formed only of joints of different lengths strung together. Upon looking, however, at the upper end of the joints you will find that each has a pair of opposite slightly prominent expansions, which stand in the room of leaves; and at the end of some of the shoots these expansions are closer together, more evident, connected with shorter joints, and altogether produce the appearance of slender cones. Still no flowers meet the eye. But above each of the joints of the cones, you may remark three minute scales, placed in such a way as to form a triangle; if with a fine pointed instrument you gently remove one of the scales, you will find below it, in a little niche, an ovary with a short ragged stigma, and one or, occasionally, two stamens. This is the flower, of which the external scale is all that remains to represent the calyx. The seed and seed-vessel are something like those of Chenopodium, only the former is hairy.

In Salicornia the ordinary structure of the order is, you perceive, interfered with, by the imperfect formation of the leaves and calyx, by the number of stamens being fewer than usual, and by the peculiar structure of the jointed stems. In some other genera the aspect of the plants is changed by a curious peculiarity in the calyx; in one plant, when the seed is ripe, that part is succulent, and richly coloured with crimson, and as the flowers grow in compact clusters, the calyxes, readily adhering, form small balls, with much the appearance of Strawberries, whence such plants are called *Strawberry Blite* (Blitum). In others, the calyx, at the angle where it bends over the seed-vessel, expands into membranous wings, giving the whole plant a very singular appearance; this occurs in *prickly Saltwort* (Salsola Kali), a common plant on the sands of the sea-coast in some parts of this country, and sometimes collected for the sake of its ashes, which yield common soda in abundance. A tendency to this enlargement of the calyx exists even in common Beet, whose seed-vessel is surrounded by the calyx, half in a hardened, half in a spongy state.

Having thus made yourself mistress of the peculiarities of the Goosefoot Tribe, let me recommend you to contrast them with the Nettle Tribe, the Buckwheat Tribe, and the Amaranth Tribe, because all those are, in reality, very closely allied to it. Parallel columns had better be again employed for comparing their differences—

NETTLE TRIBE.	BUCKWHEAT TRIBE.	AMARANTH TRIBE.	GOOSEFOOT TRIBE.
Stipules membranous and distinct.	Stipules membranous and ochreate.	Stipules o.	Stipules o.
Flowers of two sorts.	Flowers of one sort.	Flowers uniform, dry, membranous, and surrounded by bracts.	Flowers either uniform, or of two sorts, soft, succulent, and not surrounded by bracts.
Seeds round.	Seeds triangular.	Seeds round.	Seeds round.
Radicle at the point of the seed.	Radicle at the point of the seed.	Radicle at the base of the seed.	Radicle at the base of the seed.

Hence it appears that the Nettles and Buckwheats have stipules, and the radicle at the point of the seed;

while the Amaranths and Goosefoots have no stipules, and the radicle at the base of the seed. At the same time the Buckwheats differ from the Nettles in having ochreate stipules, uniform flowers, and triangular seeds; while the Goosefoots differ from the Amaranths in having herbaceous, succulent, naked flowers, and in very little else. For this reason some persons would combine the two latter Natural Orders; but they are recognized as distinct by almost all Botanists.

EXPLANATION OF PLATE XL.

I. THE BUCKWHEAT TRIBE.—1. A piece of *Knot-grass* (Polygonum aviculare).—2. A single leaf with its ochreate stipules, *a*, which are rent asunder by the expansion of the stem that they surrounded.—3. A flower.—4. The same split open, shewing seven stamens, arising from a fleshy tube of the calyx.—5. An ovary cut open, shewing the three stigmas, and the single erect ovule.—6. A ripe seed-vessel, or nut, invested by the remains of the calyx.—7. A section of the seed, shewing the embryo lying on one side of the mealy albumen.

II. THE GOOSEFOOT TRIBE.—1. A twig of *Goosefoot* (Chenopodium album).—2. A flower before unfolding.—3. An open calyx, with the stamens and ovary.—4. An ovary, with the two stigmas.—5. The same cut open, and shewing the ovule resting on its end.—6. A seed, with the torn remains of the membranous seed-vessel investing it. —7. A section of the seed, shewing the curved embryo, and the mealy albumen in its centre.

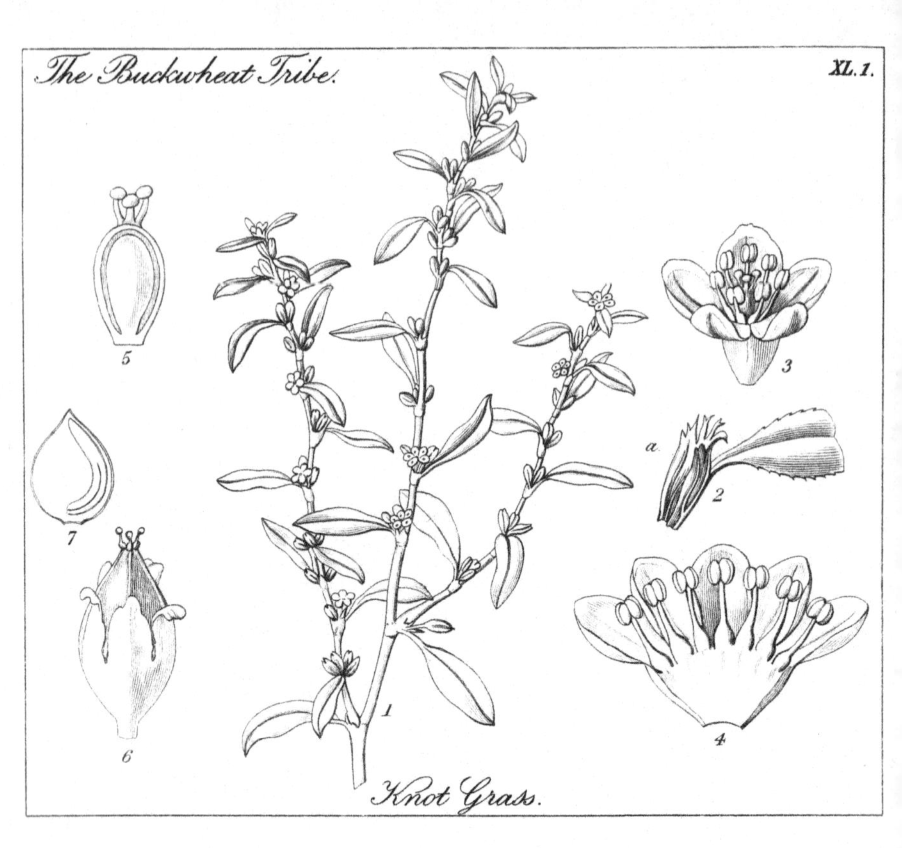
The Buckwheat Tribe.
XL. 1.
5
3
a
2
7
6
1
4
Knot Grass.

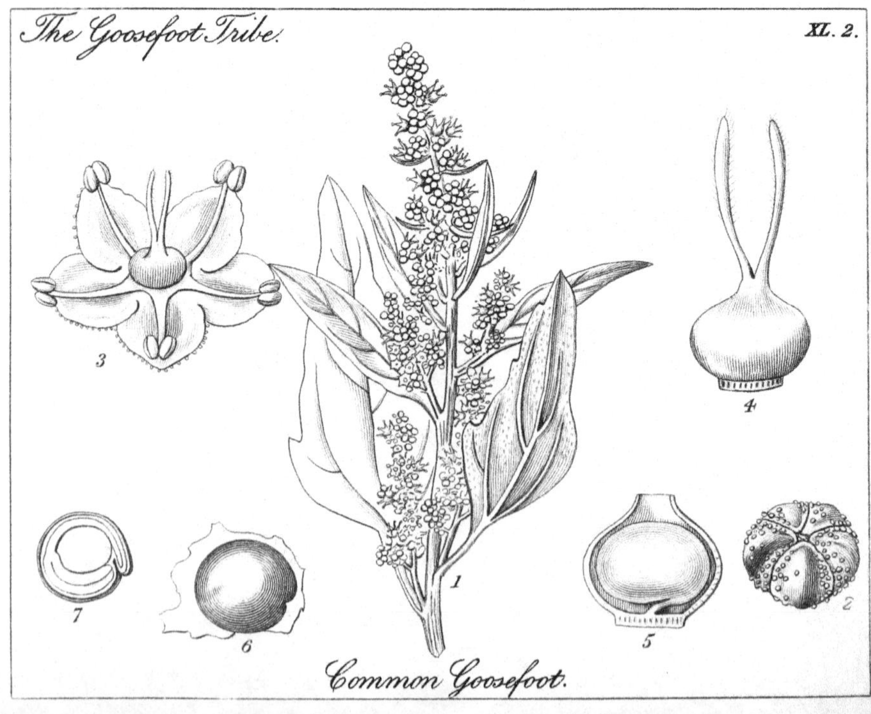
The Goosefoot Tribe.
XL. 2.
3
4
7
6
1
5
2
Common Goosefoot.

LETTER XL.

THE MEZEREUM TRIBE—THE CINNAMON TRIBE.

(Plate XLI.)

It is by no means an unusual thing for the prettiest and most splendid species of the vegetable kingdom to conceal the deadliest qualities. Rhododendrons, Azaleas, and Kalmias, in the Heath Tribe, Blood-flowers (Hæmanthus) and Crinums in the Narcissus Tribe, Foxglove in its own tribe, and Ranunculuses, Aconites, and Larkspurs, in that of the Crowfoot are familiar instances of this; to which you now may add that of the various pretty species of the Mezereum Tribe. The plant from which the name is derived (Daphne Mezereum), the *Spurge Laurel* (Daphne Laureola), the lovely *trailing Cneorum* (Daphne Cneorum), and various other species of the same genus, together with the Gnidias, and Struthiolas of the greenhouse, are all acrid, suspicious plants, and in some instances extremely dangerous. The berries of the Mezereum and the Spurge Laurel are fatal poisons to man, although birds feed upon them uninjured; the bark of all the species is so acrid, that if moistened and bound down upon the skin it raises blisters, and even the perfume of those which have fragrant flowers will often produce fainting in persons with delicate nerves.

The ingenuity of the fair sex has not failed to profit by these qualities, for the Tartar ladies, availing themselves of the acrid property of the Daphne leaves, rub them over their cheeks instead of rouge, to raise a gentle colour in the skin.

All this while I am talking of these plants as if you knew them, and I think it impossible but some of them must have been already seen by you. Spurge Laurel is a common evergreen in shrubberies, with deep green, shining leaves, little pale green flowers, almost concealed by the leaves, and black drupes resembling those of the common Laurel (Prunus Laurocerasus) externally.

Still more common is Mezereum, and much more striking, for it bears its rose-coloured or white flowers upon naked branches late in autumn or early in the spring, and at that time there is no shrub that at all resembles it. We will take this species for examination. The rosy flowers do not owe their colour to the presence of petals, but are merely composed of a calyx (*Plate* XLI. 1. *fig.* 2.), having four spreading lobes half-united in a tube, hairy externally and slightly pitted all over the inner surface. This arises from the looseness of the parenchyma which connects the two surfaces of the calyx, and which is so easily disturbed, that you may without difficulty separate the whole of the inner from the whole of the outer surface, dividing the calyx into two cups. Hence it has been thought by some, that in reality the calyx and corolla are really both present in Daphne, but naturally glued together. Such an opinion is however necessarily unfounded, as will be obvious to you, if you call to mind the rule I sometime since explained to you (page 131), that all the parts of a flower naturally *alternate* with each other. Nevertheless, M. Dunal, a French Botanist, has founded upon this and some other cases, a theory of *unlining* in flowers (*dédoublement*), imagining that in all cases the corolla is produced by an unlining of the calyx!

But to return to our Mezereum. Eight stamens in two rows, one above the other, are placed on the tube of the calyx; and at the bottom of the cup is a superior one-celled ovary (*fig.* 2.), with a nearly sessile, tufted stigma (*fig.* 3.). In the inside of the ovary hangs a single ovule (*fig.* 3. *c.*), with a foramen (page 73, and *fig.* 3. *b.*), so conspicuous, that it may be almost seen by the naked eye. When you first pull the ovary of the Mezereum in pieces you will probably imagine that the ovule is enveloped in a loose hairy bag (*fig.* 3. *a.*); but upon scrutinizing it more narrowly, you will find that the supposed bag is merely the lining of the ovary, which readily separates from the shell and clings more or less to the ovule; so that you see the disposition to *unline*, which is found in the calyx, is also conspicuous in the ovary. A section of the ovary, carefully made in a vertical direction (*fig.* 3.), exhibits this very clearly, presenting the appearance of two cavities, one above the other, the upper one smooth and containing the ovule, the lower one bristling with crystalline points, and empty.

The ripe fruit of Mezereum is red and succulent

(*fig.* 4.); it has a slight depression at the point where the stigma was, and it contains a single suspended seed (*fig.* 5.). The embryo is a large almond-like (*amygdaloid*) kernel, with two fleshy plano-convex cotyledons, and a radicle pointing to the apex of the fruit (*fig.* 6.).

All the genera you are likely to meet with, belonging to the Mezereum Tribe, are very like the above n nearly every respect; the most important differences to remember, are, that some of them, as Gnidia, have little scales in the mouth of the calyx, and that it is only in a few that the fruit is succulent: more generally it is hard, dry, and nut-like.

You have before had instances of the toughness of the bark of plants, and of its fitness to be manufactured into cordage or similar materials. All the Daphnes, and indeed the Tribe, partake of this quality; they are, however, chiefly known for the production of paper and lace. In China and India, a coarse paper is manufactured from the inner bark of two or three species; and in the West Indies, Nature produces a beautiful description of lace in that of another kind. You seem incredulous; and yet I know not wherefore you should be, considering how many greater wonders you have already witnessed among plants. I repeat it, in Jamaica a natural lace, of fine quality, is produced ready manufactured, in the bark of the Lagetto Tree; so fine, indeed, that ruffles, a frill, and cravat were cut from it, and sent as a present to King Charles II. This substance is not so often seen in England as its beauty and curiosity would have led one to expect; in Jamaica it is well known, and, by a little contrivance, the most beautiful brushes are readily prepared from it. In reality it consists of the thin layers of the inner bark of the tree, a little stretched sideways, so as to separate the parallel fibres, and to give their meshes a lozenge form. Botanically considered, it is the very same substance as that from which are manufactured the common Russia Mats, in which furniture is packed, and of which Gardeners make use under the name of *Bast;* only instead of being coarse and brown, it is extremely fine and white.

It is not a little curious, that of the reputed Laurels so very common in gardens, namely, the Common Laurel, the Portugal Laurel, the Spurge Laurel, the Alexandrian Laurel, the two first should be Cherries, the third a Daphne, the fourth a Ruscus, and not one of them in truth a Laurel; while, on the contrary, a species of true Laurel, actually cultivated very commonly, is not recognised popularly, but has the name of *Sweet Bay*. For this reason, although it would have been more correct to have called the assemblage of plants to which the latter belongs the Laurel Tribe, yet, to avoid confusion, it is better to drop the name Laurel altogether, and to designate the plants by a title about which there can be no mistake. We will, therefore, call the Lauraceous plants of Botanists the Cinnamon Tribe, because the aromatic spice of that name is yielded by some of the species, and because an aromatic principle, of an analogous kind, is found throughout the remainder.

These useful and valuable plants are most curious in their structure, as the *Sweet Bay* (Laurus nobilis) will shew you. It is, as you are aware, a hardy evergreen bush, whose leaves, when bruised, give out a grateful spicy odour, and which are in consequence extensively used in making pastiles and sweetmeats. It grows wild in the south and middle of Europe, especially in France and Italy, where it is a great ornament of scenery.

The flowers break out of little brown scaly balls, in the axils of the leaves (*Plate* XLI. 2. *fig.* 1.), are of a pale cream colour, and have the following structure. The calyx is divided into four deep lobes, which spring from the top of a hairy stalk (*fig.* 2. & 4.). There are no petals. Some flowers are sterile, some fertile. The sterile flowers (*fig.* 2.) contain eight stamens of a most singular nature; each has a flat, linear filament, with a stalked, kidney-shaped gland, growing from each side near the base (*fig.* 3. *c.*), and a two-celled anther, the valves of which separate from their cells, like those of the Barberry (page 12.), and curve back by a sort of hinge at the upper end of each cell. It is found by the structure of other genera, that the kidney-shaped glands are abortive stamens; consequently, in this apparently simple flower, there is an irregularity of a very extensive kind, four petals and sixteen stamens being either abortive or rudimentary, as is expressed by the following scheme.*

* See page 132.

S S S S
. . . .
s s s s
s s s s
s *s* *s* *s*
s *s* *s* *s*

In the fertile flowers (*fig.* 4.) there are only four imperfect stamens, each with two, small, rounded lobes in the middle, and a superior ovary containing a single cell, and terminated by a style surmounted by a dilated, purple, glandular stigma (*fig.* 5.); it contains a single suspended ovule. The fruit, when ripe, is ovate, black, and succulent (*fig.* 6.), and contains a single seed, with a thick, almond-shaped (*amygdaloid*) embryo, whose radicle points to the apex of the fruit (*fig.* 7. & 8.).

It will be plain to you at once that this plant corresponds in every respect with the Mezereum Tribe, as in the absence of petals, the number of lobes of the calyx, the number of perfect stamens, and the whole structure of the ovary and fruit. But, on the other hand, it differs in the presence of so many abortive stamens, in its anthers with recurved valves, and its aromatic qualities. It is, however, well worth notice, that although we use Cinnamon and Cassia habitually in small quantities, yet that this Tribe possesses in abundance a powerful acrid juice, which assimilates its secretions very much to those of the Mezereum Tribe.

The genera differ chiefly in the number, form, and position of the abortive stamens, in the number of cells of the anther, in the form, &c. of the calyx, and some other minor circumstances. They are, however, so seldom seen alive in Europe, in a flowering state, that I need not occupy your time about them. Consider them aromatic Mezereum-like plants, with anthers bursting by recurved valves, and you cannot easily forget them.

EXPLANATION OF PLATE XLI.

I. The Mezereum Tribe.—1. A twig of *common Mezereum* (Daphne Mezereum).—2. One of its calyxes cut open, and shewing the stamens and ovary.—3. An ovary cut perpendicularly, to shew the cavity *a* in the pericarp, the ovule *c*, and its foramen *b*.—4.* and 5.* stamens.—4. A cluster of ripe fruit, natural size.—5. A section of a fruit, shewing the seed.—6. The embryo.

II. The Cinnamon Tribe.—1. The *Sweet Bay* (Laurus nobilis).—2. A sterile flower, with its 8 glandular stamens.—3. A stamen separated; *a* one of the valves of the anther, *b* another quite rolled back, *c* abortive stamens, resembling glands.—4. A fertile flower, with the four abortive stamens surrounding the ovary.—5. A section of the ovary.—6. A ripe fruit.—7. A longitudinal section of the same, shewing the radicle at the apex.—8. A transverse section of the same, shewing the two cotyledons cut across.

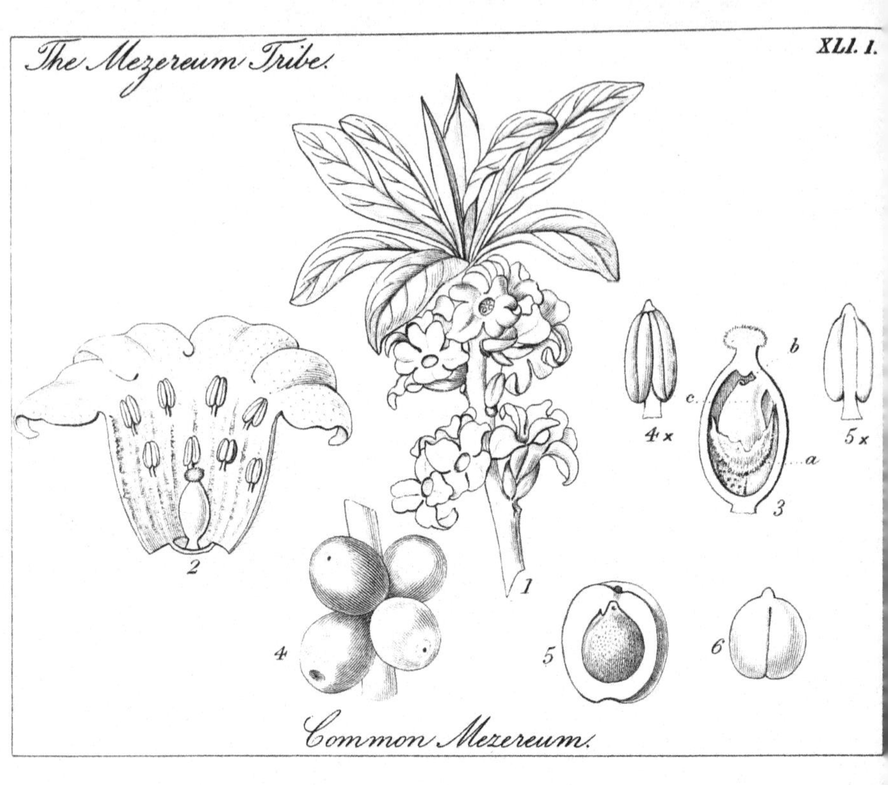
The Mezereum Tribe.
XLI. 1.
1
2
3
a
b
c
4
4 x
5
5 x
6
Common Mezereum.

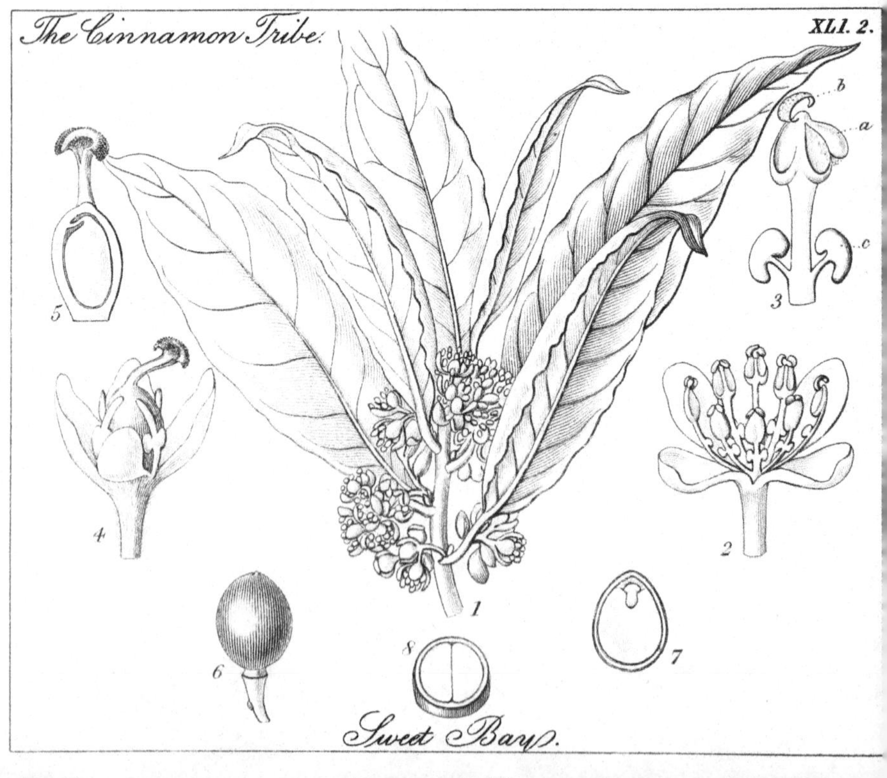
The Cinnamon Tribe.
XLI. 2.
1
2
3
a
b
c
4
5
6
7
8
Sweet Bay.

LETTER XLI.

THE PRIMROSE TRIBE—THE EPACRIS TRIBE.

Plate XLII.

The Primerose, *Primrose,* or First Rose of Spring, the *Cowslip,* the *Oxlip,* the *Auricula,* or Powdered Beau, are so associated with our earliest recollections as children, that we never, to the last hour of our existence, entirely cease to look upon them with pleasure. Nor indeed is it possible, independently of all our remembrances of infancy, to behold without delight a sunny bank all light and life with tufts of sweet yellow flowers, when nature elsewhere remains in the garb of death and winter.

All these pretty plants are so well known to you, that I might as well have left you to study them by themselves, without any remark beyond what has already been made upon a former occasion (Vol. I. p. 187.), if it were not my wish to illustrate in detail every natural assemblage that is common enough to be likely to fall in your way. Leaving you then to refer to our correspondence for the general facts connected with the Primrose Tribe, let us carefully note down the structure of the Cowslip blossom *(Plate* XLII. 1.).

You know that the flowers grow in little nodding clusters, or umbels, from the top of a round, brittle stalk, which raises them above the low herbage that surrounds them. Each has a tubular, angular, pale green calyx, with five teeth corresponding with the projecting angles. The corolla (*fig.* 2.) is monopetalous, salver-shaped (that is with a cylindrical tube and flat border), and a little contracted near the middle of its tube; its border is divided into five lobes, each of which is slit or notched at the end. The anthers are almost without filaments, and stand side by side half way down the tube of the corolla; not, however, alternately with its lobes, as usually happens, but opposite to them. Perhaps you will not very readily ascertain this without assistance; either of two modes will enable you to do so. Look down into the tube without disturbing any of the parts, and you will then see that each stamen stands in front of a lobe of the corolla; or, if this does not satisfy you, cut open the corolla, hold it against the light, and you will then see that each anther stands upon a delicate, half-transparent vein, which passes through the middle of a lobe of the corolla (*fig.* 2.). This is an important fact, for two reasons. In the first place, it enables you to distinguish a plant of the Primrose Tribe with certainty from all other monopetalous tribes, except one; and secondly, it indicates a great degree of irregularity in the flower; for by the rule I have lately given you (page 132.) it appears, that, simple and unchanged as the Cowslip flower apparently is, it really has five of its stamens, namely, the first row, absent, those which make their appearance belonging to the second row.

The only other monopetalous tribe in which the stamens are opposite the lobes of the corolla, is the Myrsinaceous, to which belong the Ardisias, so frequently seen in stoves; but all these are exotics, and are bushes or trees, by which circumstance they are at once known from the Primrose Tribe, without taking other circumstances into account.

The ovary of the Cowslip (*fig.* 5.) is a superior hollow case, of a top-shaped form, and marked by ten longitudinal furrows, of which five are more conspicuous than the others; it is surmounted by a slender thread-shaped style, which terminates in a round pin-headed stigma. From the number of furrows on the outside of the ovary you would naturally expect that it should contain five cells; but upon opening it you can only discern a single cavity, with a free central placenta (*fig.* 6. *a.*), loaded with ovules in the middle. If, however, you were to dissect the ovary, when very young, you would then find five cells cut off from each other by five partitions; but long before the flower opens these partitions are broken and disappear, and the consequence is that the placenta grows up in the middle without any connection with the sides of the ovary.

The fruit of the Cowslip is well worth an attentive study. It is invested with the dry, dead, but firm and little altered calyx (*fig.* 7.), at the bottom of which stands a seed-vessel that splits at the point, with ten teeth which turn backwards to allow the seeds to fall out; its centre is occupied by a brown hard cone, over which the seeds are closely arranged. Each seed is a little, rough, deep-brown, angular body (*fig.* 8.), adhering to the placenta by the middle; and containing a dicotyledonous embryo lying across the hilum (*fig.* 9.); that is, not turning to the hilum either the radicle or the end of the cotyledons.

The genera of the Primrose Tribe are very distinctly marked. Cyclamen or *Sowbread*, for instance, has the lobes of its corolla bent back, and when the flower is past it gently twists its peduncle till it becomes so short as to bury the tough leathery seed-vessel in the earth. This is a most curious economy, the cause or object of which is quite unknown. Anagallis, or the *Pimpernel*, has the corolla divided into five deep lobes, and its seed-vessel opens by a lid like a little soap-box. Hottonia or *Water-violet*, with its feathery leaves, and beautiful pink flowers, has a five-parted calyx, the corolla of a Primrose, and a round seed-vessel, with five compact valves. *Brookweed* (Samolus Valerandi), with its tassels of small white flowers, has a partially inferior ovary, and five additional rudimentary stamens in the form of tapering threads. And the remainder are equally easily recognized.

And now you know as much about the Primrose Tribe, in general, as it is necessary for me to tell you in this correspondence; for the rest you will, of course, refer to systematic works. I must, however, before I quit the subject, tell you of one delicate, tender, little, lovely thing, that is seldom seen, except by Botanists, but which you might easily cultivate in your window, under a bell-glass. This is the Bog Pimpernel; a small, trailing, light green plant, with stamens only a few inches long, and covered with tiny roundish leaves, most beautifully dotted on their underside. The flowers of this plant are little rose-coloured bells, standing upright upon slender stalks, which afterwards curve downwards to bury the fruit. Its stamens are so thickly covered with the most delicate, glittering, jointed, entangled hairs, that they look like five yellow anthers, standing on the top of a tuft of silver wool. Where the Sun-dew, Parnassia, Butterwort, and all those curious little bog plants occur, there you will be sure to meet with this charming species, which Botanists have truly called the *delicate* (Anagallis tenella), for we have no more fragile flower in all our British Flora.

From these we turn to plants so like the Heath Tribe (Vol. 1. page 158. *Plate* XII. 1.) that it is useless for me to do more than point out in what they differ. I am adverting to the Epacris Tribe. In the smallness of their leaves, the brittleness of their stems, their uses to man, and the gayness of their flowers, in the clearness and brilliancy of their colours, and in general points of organization, in short, in all that can strike the casual observer, the two tribes are alike; one being the pride of Europe and Africa, the other the glory of the hills and wastes of the Australian continent and islands. Several species are common in greenhouses, particularly of the genus *Epacris*, with their bells of pink, or red, or white, of *Styphelias* and *Dracophyllums*, with their long tubes of crimson and other colours, or of *White beards* (Leucopogons), with the mouths of their corolla stuffed with snow-white hairs. The *Pink Epacris* (Epacris ruscifolia, *Plate* XLII. 2. *fig.* 1.) is one of the easiest to procure.

It forms a slender heath-like plant, with stiff, ovate, sharp-pointed leaves, in the bosoms of the uppermost of which the flowers are closely arranged upon short stalks, covered completely by scale-like bracts (*fig.* 2.), which one can hardly distinguish from the five-leaved calyx. The tubular bell-shaped corolla, with a short, spreading, five-lobed border, the superior ovary with five many-seeded cells (*fig.* 6.), the hypogynous scales, (*fig.* 5. *a.*), the single style, the obscurely lobed stigma, and, finally, the dry seed-vessel, containing a vast quantity of minute seeds (*fig.* 8.), are all characters in accordance with those of the Heath Tribe. But in the stamens there is a material difference; they arise from the side of the corolla (*fig.* 3.) in this species, but sometimes they agree with the Heath Tribe in growing from below the ovary: the anthers are always one-celled, opening by two valves (*fig.* 4. *a.*), and never two-celled, opening by five valves, or two pores. It must be confessed, these are not points of much importance, but Botanists seem generally agreed upon recognizing the Epacris Tribe as distinct from that of Heaths, and we may as well swim with the stream, as it is really of no practical importance whether the Epacris Tribe is considered a distinct assemblage, or a mere section of the Heath Tribe.

In general, both the Epacris and Heath Tribes have a dry seed-vessel; but as we have the succulent Arbutus and Cranberry among the latter, so we have the *Australian Cranberry* (Lissanthe sapida and Astroloma humifusum), and *Van Diemen's Island Currants* (various Leucopogons) among the former.

EXPLANATION OF PLATE XLII.

I. THE PRIMROSE TRIBE.—1. A cluster of *Common Cowslip* (Primula veris) flowers.—2. A corolla opened to shew the position of the stamens, with respect to the lobes of the corolla.—3. A calyx with a portion cut away to shew the ovary, style, and stigma.—4. A stamen.—5. A pistil very much magnified; in this figure the furrows of the ovary, and the glandular hairs of the style, are distinctly seen.—6. A cross section of an ovary, with the free central placenta at *a*.—7. Half a seed-vessel, invested with the calyx; *a a* are the teeth-like valves by which it opens, and *b* the central free placenta covered with seeds.—8. A ripe seed, much magnified; *a* the hilum.—9. A section of the same shewing the embryo lying in hard albumen ; *a* the hilum.

II. THE EPACRIS TRIBE.—1. *Pink Epacris* (Epacris ruscifolia).—2. A flower with the scale-like bracts that cover its stalk.—3. A corolla cut open, shewing the stamens, ovary, and style.—4. A one-celled anther opening by two granular valves, *a a*.—5. A pistil, with five scales *a*, surrounding the base of the ovary.—6. A cross section of the ovary, shewing the five cells, and many-seeded placentæ.—7 A fruit of Epacris (after Gærtner).—8. A section of it shewing the placentæ and seeds.—9. A seed cut in half lengthwise, shewing the embryo lying in albumen, with the radicle turned to the hilum.

LETTER XLII.

THE GREEK VALERIAN TRIBE—THE TRUMPET-FLOWER TRIBE.

(Plate XLIII.)

WHEN we were studying the Bindweed Tribe (Vol. I. *p.* 161.), I neglected to mention a set of common plants, closely allied to it, but in general not twiners; namely, the GREEK VALERIAN TRIBE, or Polemoniaceous Plants. Let me now proceed to supply that omission.

You cannot but be acquainted with the genus *Phlox*, or, as the old gardeners called it, Lychnidea, the species of which are among the showiest of common perennials, whether they rise erect from the ground with broad, deep-green, opposite leaves, and dense clusters of purple flowers, terminating the branches, or lie prostrate among rock-work, with their slender stems covered by sharp scales, and a few neat lilac or white blossoms just raising themselves from the soil.

Cobæa is another well known genus, that overruns the forests of Mexico with its annual stems, and which, on that account, has been long since transferred to the gardens of this country, where it will grow between two and three hundred feet in a single summer, completely hiding a large extent of treillage with its

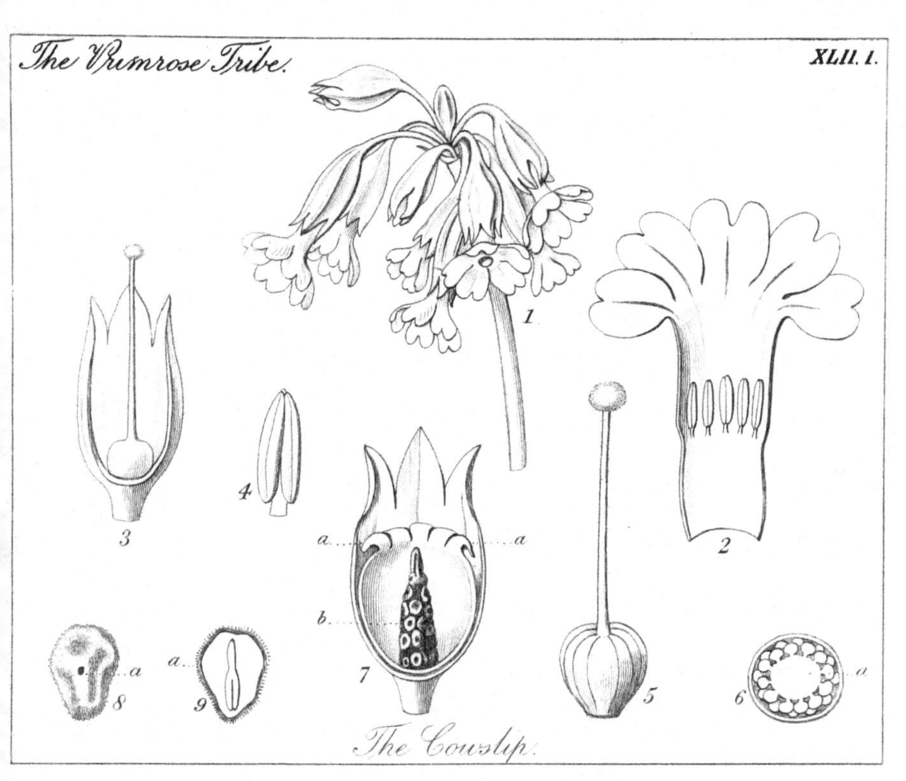

The Primrose Tribe.
XLII. 1.
1
2
3
4
5
6
7
8
9
a
b
The Cowslip.

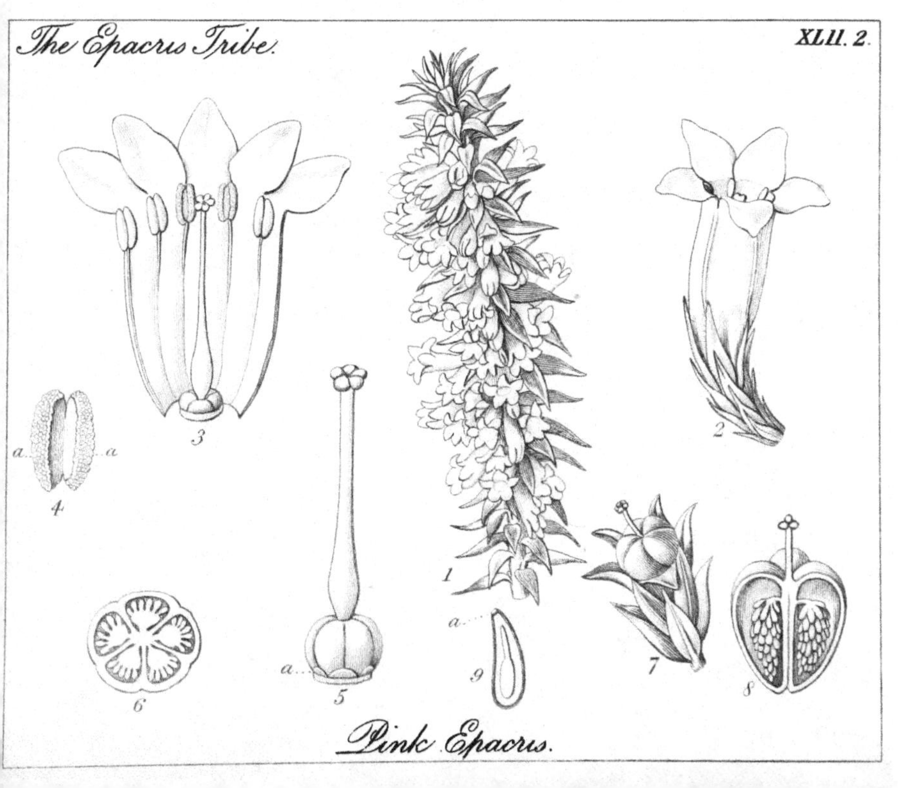

The Epacris Tribe.
XLII. 2.
1
2
3
4
5
6
7
8
9
a
Pink Epacris.

blueish green leaves, and huge, bell-shaped, dingy, greenish-purple blossoms.

With these, Botanists associate the *Gilias*, with their heads of blue, or red, or party-coloured flowers, and finely cut leaves, the *Collomias* with their small buff or brick-red blossoms, peeping from among close glandular bracts, *Ipomopsis* with its innumerable pendent tubes of scarlet and gold, and the *Greek Valerians* (Polemonium), or *Jacob's Ladder* plants, the old-fashioned gardeners' pets, with their spreading fern-like leaves, and nodding bells of blue or white.

We will, however, study none of these. Let us rather examine a plant of this charming annual lately imported by the Horticultural Society from California (*Plate* XLIII. 1.); it is called *Many-coloured Slender-tube* (Leptosiphon androsaceus). Observe how delicately it is frosted by little glandular hairs; millions of millions of these bodies must be perpetually employed in separating from the blood of the Slender-tube the matter which Nature requires it to part with. Its leaves are divided into deep narrow lobes, which all spring from near the same point. Its flowers stand in an umbel, at the end of a slender stalk, and have their bases buried among narrow green bracts. Each calyx (*fig.* 3.) has five, narrow, sharp-pointed, hairy lobes, connected into a short tube by a thin web (*fig.* 3. *a.*). The corolla has a slender, reddish-brown tube, with a spreading, five-lobed, pale, lilac border, yellow at the base, and within the tube deep-chocolate brown (*fig.* 2.). It has five anthers, stationed on short filaments at the orifice of the tube.

and projecting a little way beyond it. The ovary is superior, contains three cells (*fig.* 5.), in each of which are about six ovules adhering to a placenta in the axis. The style is thread-shaped, and terminated by three narrow lobes (*fig.* 3.). The seed-vessel (*fig.* 6.). opens into three valves, bearing the dissepiments in their middle. The seeds are spongy, oval bodies (*fig.* 7.), with a thick skin, and contain an erect embryo (*fig.* 8.) without albumen.

If you contrast this with a Bindweed, you will remark that in that plant, the corolla has its lobes plaited together, the stigma two-lobed, more or less, while here the lobes of the corolla are imbricated, and the stigma three-lobed. These distinctions are the most material for separating the two tribes, for we cannot make great use of the twining habit of the Bindweeds, first, because *Cobœa*, which is of the tribe now under consideration, also twines, and, secondly, because many Bindweeds do not twine.

I have already adverted to the existence of a genus called Collomia; it consists of species of little beauty; but in one of them, Collomia linearis, the microscope reveals one of the most marvellous phenomena I am acquainted with. The seeds of this plant are small, dry, hard and brown. If you look at them ever so carefully while dry, you will find nothing that can lead you to suspect the existence, in their skin, of any peculiar mechanism. But place them under a microscope, and, while watching them, gently float them in water. In a few moments the fluid will appear in rapid motion, thousands of silvery threads will be seen lancing themselves into the water, and unrolling in all directions, and the whole field of the microscope will, on a sudden, present a spectacle of action, life, and movement. This is owing to the expansion of a vast quantity of spiral threads, which, when dry, are contracted and glued to the surface of the seed, but which are suddenly set at liberty upon the application of water.

Another set of plants, that I must bring you acquainted with, is the TRUMPET-FLOWER TRIBE. You have long since studied the Foxglove Tribe, and you remember that it consists of herbs, with an erect habit, and little angular or round seeds, the embryo of which is surrounded by albumen. Very nearly akin to these are certain exotic plants, most of which are trees or shrubs, with flowers like those of the Foxgloves in all respects, only that they are usually larger, in most instances with a twining or climbing mode of growth, and with large flat pods, some of which are as much as two or three feet long, filled with flat thin-winged seeds, containing an embryo without albumen.

The common genus of these plants is named Bignonia, or the *Trumpet-flower*, whence the Tribe has gained its usual designation. All that belong to it are climbing plants, as is indeed every species common in gardens, except the noble *Catalpa* (Catalpa syringifolia), which forms a tree as large as an apple tree, and almost as hardy, its boughs loaded in summer with heaps of magnificent white and lilac flowers.

The other trees of the Tribe are, with the exception of Jacaranda, with their airy, graceful, fern-like foliage, unknown in Europe in a living state; they inhabit the forests of India and America.

The *Rooting Trumpet-flower* (Bignonia radicans), is so very common, that there is hardly a village in England where some garden does not contain it. We will, therefore, select it for study. Mr. Elliot tells us, that it is common in the damp rich soil of Carolina, "climbing over buildings and the loftiest trees, throwing out radicles all along the stem, by which it attaches itself firmly to walls, fences, and the bark of trees." In this country it is much less vigorous, owing no doubt to the greater coldness and dryness of our climate. It has opposite pinnated leaves, the leaflets of which are ovate, taper-pointed, and sawed (*Plate* XLIII. 2. *fig.* 1.). Its flowers, of the richest brown-red or blood-red, and of a fleshy consistence, grow in clusters from the ends of short stiff peduncles. The calyx is a fleshy cup, divided into five sharp, somewhat triangular teeth. The corolla is funnel-shaped, between two and three inches long, with a border divided into five roundish, rather unequal lobes. Five stamens spring from the tube of the corolla; of these, two are longer than two others, all four being furnished with diverging sharp-pointed anthers, and the fifth (*fig.* 2. *a.*) is merely a rudimentary tooth, analogous to what you find in the flower of a Pentstemon. The ovary is seated upon a thick, yellow, fleshy cushion or disk (*fig.* 3. *a.*), and consists of two cells, containing many ovules spread over the surface of a central placenta (*fig.* 4.). The ovary gradually tapers into a stiff, curved style, ending in a stigma composed of two thin plates.

Thus far the Bignonia is so like a plant of the Foxglove Tribe, that no Botanist can point out a distinction. It is otherwise with the fruit; in this species it is described as a very long tapering pod, filled with winged seeds; in other species its seeds are as follows: a somewhat wedge-shaped, rounded, flat centre (*fig.* 5.), comprehending a two-lobed embryo, without albumen (*fig.* 6.), is surrounded by a thin, delicate membrane, or wing, the whole substance of which consists of small, semi-transparent cells, round the sides of which is twisted a spiral, silver thread. It is here that the great difference between the Tribes of Foxglove and Trumpet-flower resides. The former has no wing to its seeds, nor any thing like the form of a long pod in its fruit. Remembering this then, you never need confound the one with the other.

Eccremocarpus scaber, is one of the prettiest of the Bignonia Tribe. From the hedges and thickets about Valparaiso, it has been transferred to our gardens, where it survives moderate winters without injury. The curious rough pods of this plant produce an abundance of the winged seeds of the Bignonias, and are well worth a careful examination.

EXPLANATION OF PLATE XLIII.

I. THE GREEK VALERIAN TRIBE.—1. A cluster of the flowers of the *Many-coloured Slender-tube* (Leptosiphon androsaceus).—2. The tube of a corolla cut open, to shew the origin of the stamens.—3. The calyx, style, and stigmas; *a* the membranous web that connects the lobes.—4. A longitudinal section of an ovary.—5. A transverse section of the same.—6. A ripe capsule, much magnified.—7. A seed.—8. The same cut through longitudinally, to show the embryo.

II. THE TRUMPET-FLOWER TRIBE.—1. A cluster of flowers of the *rooting Trumpet-flower* (Bignonia radicans).—2. A portion of the tube of the corolla, cut open to shew the origin of the stamens; *a* the fifth rudimentary stamen.—3. The pistil; *a* its disk, *b* the ovary.—4. A transverse section of the ovary.—5. A ripe seed of Bignonia indica.—6. Its embryo; *a* the two-lobed cotyledons, *b* the radicle.

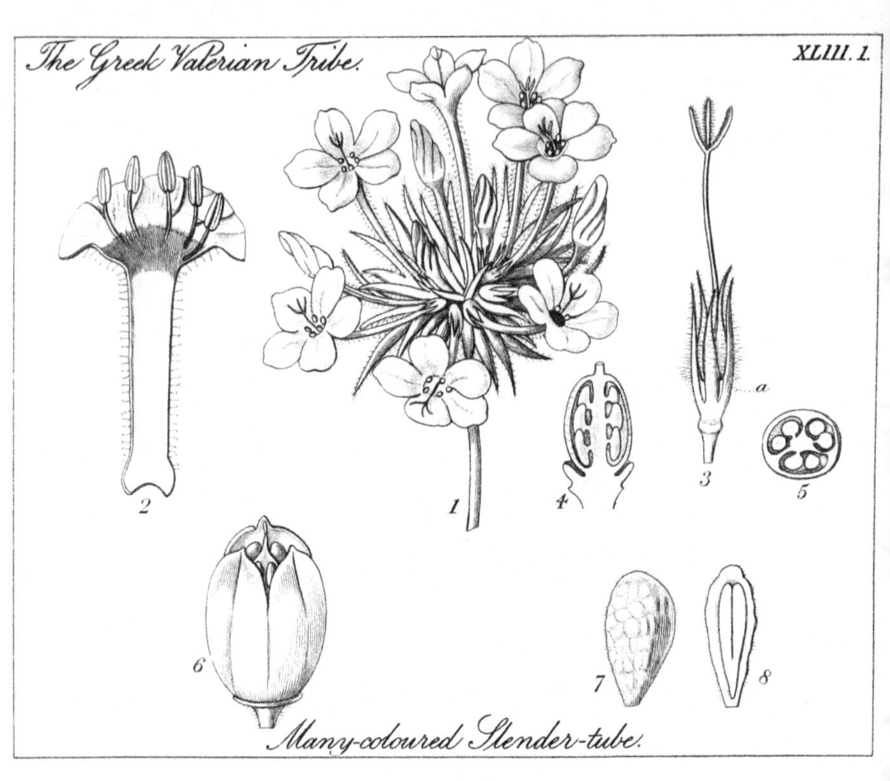
The Greek Valerian Tribe.
XLIII. 1.
a
1
2
3
4
5
6
7
8
Many-coloured Slender-tube.

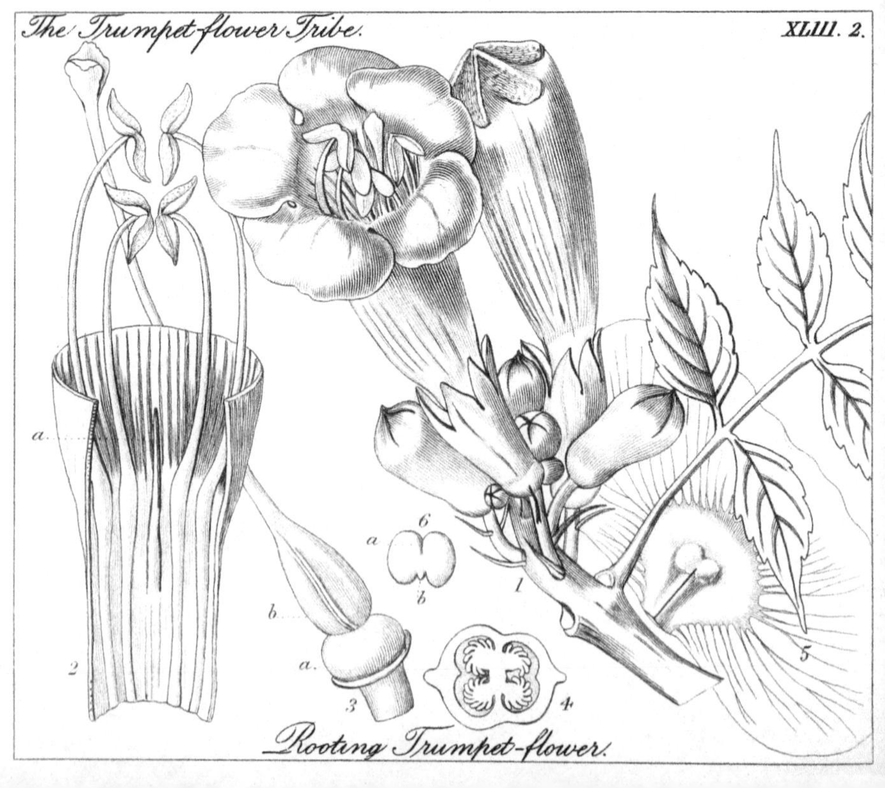
The Trumpet-flower Tribe.
XLIII. 2.
a
a
6
b
b
a.
1
2
3
4
5
Rooting Trumpet-flower.

LETTER XLIII

THE MADDER TRIBE—THE SCABIOUS TRIBE.

Plate XLIV.

YOUR old housekeeper is right; on this occasion she proves a better Botanist than her mistress. There is no danger whatever in using Goosegrass for a sieve; and it is quite true that it was formerly so employed, until the cheapness of wire-work superseded it. Only take a handful of the dry stems, press them into a bowl without a bottom, or into a colander, and pour over them any thing you have a mind to strain. You will see that the liquid will come away as clear as if it had been poured through a sieve. Goosegrass possesses this property in consequence of its surface being covered over with myriads of fine hairs, hardened and curved by the hand of nature into hooks (*Plate* XLIV. 1. *fig.* 4. & 7.), which catch up and hold fast whatever may float in water, with just as much certainty as it would be intercepted by the close meshes of a sieve. As to its harmless qualities, you need have no fear upon that score, if you remark its near affinity to the Honeysuckle and Coffee Tribes (Vol. I. *p.* 176.). Let me just note down the principal points that are worthy of notice in its structure.

Goosegrass, Cleavers, Whiptongue (Galium Aparine),

or whatever else its name may be, is a herbaceous plant, usually growing in hedges and dry ditches, where its long, angular, brittle stems can readily find something to catch hold of, so as to be supported a little above the earth. The angles of its stem, the upper surface and edges of its leaves, and its fruit, are closely beset with the hard, stiff, hooked hairs, above referred to; they catch hold of the clothes of the passer-by, and adhere to him like a bur, on which account the Greeks used to call this plant the *Philanthropist.* The leaves are arranged six, seven, or eight, in a whorl; they are of a narrow figure, somewhat broader towards the upper end, terminate in a hard spiny point (*fig.* 7. *a.*), and have no stipules. The flowers grow from the bosoms of the uppermost leaves of the branchlets, in the form of tiny, white, four-rayed stars. The calyx is the slightest little edge that you can imagine, placed on the top of a small hairy ball, which is the ovary. The corolla (*fig.* 1.) is valvate and monopetalous, but almost divided into four parts, so slight is the degree of connection between the petals. From the recesses of the corolla spring four stamens (*fig.* 1. & 2.). The ovary is a round, inferior, hairy ball, containing two cells, in both of which is one ovule, rising up from a very short stalk (*fig.* 2.); it is surmounted by two styles, each bearing a single round stigma. On the outside of the styles, between them and the corolla, is a green, fleshy, two-lobed disk (*fig.* 2. *a.*). The fruit consists of a pair of kidney-shaped achænia, or nuts, bristly all over with stiff hooks, separated at the base by the hardened and widened axis, and curving inwards till their points nearly meet (*fig.* 4.). In the inside of each nut is a curved seed, containing a small embryo of the same figure, embedded in hard horny albumen, and turning its radicle towards the base (*fig.* 5.).

You cannot have a better example than this of the great importance of botanical knowledge, in forming a correct opinion upon many common questions. A person, unacquainted with the science, would not comprehend the possibility of this Goosegrass being allied to the Honeysuckle (Vol. I. *Plate* XIV.), and yet I shall shew you, by the plainest evidence, that such is an indisputable fact.

Gather a specimen of any common Honeysuckle, and compare it with another of the Snowberry, which Linnæus used to consider a sort of Honeysuckle; then place by the side of the Snowberry a Laurustinus in flower, and by the Laurustinus a bunch of Elder blossoms. You will then find, although the Honeysuckle and the Elder at first seemed very dissimilar, yet that the two may be gradually connected by so few as these two transitions.

Next, compare the Goosegrass with the Elder. The former has a small, white, regular, monopetalous corolla, with as many stamens as lobes, an inferior ovary, containing one seed in each cell, seeds with an embryo buried in horny albumen, and opposite leaves without stipules; in all these important points the Elder coincides. That plant, indeed, is a small tree, with pinnated leaves, large cymes of flowers, three cells to the ovary, and succulent fruit, while Goosegrass is a prostrate, annual, rough-stemmed plant, with simple whorled leaves, solitary flowers, two cells to the ovary, and dry bur-like fruit. But such matters are irrelevant to the discussion; for as there is no question as to the great difference of these two plants, the point to determine is, whether they are related to each other, and, if so, in what degree. It is impossible to deny that the points of coincidence which I have named to you, are sufficient to establish the fact of their relationship; and, therefore, as it is proved that the Goosegrass is related to the Elder, and the Elder to the Honeysuckle, it follows that the Goosegrass and the Honeysuckle are also related to each other; but not equally. The Elder and the Honeysuckle are plants of the same natural Order (or Tribe, as it has pleased us to call natural Orders in this correspondence), and may therefore be said to be related to each other in the first degree; Goosegrass, on the contrary, belongs to a distinct natural Order, and therefore cannot be related in more than the second degree.

Goosegrass may, in fact, be taken as the type of the MADDER TRIBE, the peculiar distinctions of which are drawn from the angular stems, whorled leaves, and double one-seeded ovary. *Madder* itself (Rubia tinctorum), from the roots of which a valuable dye is extracted, is very much like a Galium, but is more vigorous in its mode of growth, has larger hooks, and a succulent fruit; in which latter respect a greater approach is made to the Elder than in the case of Bedstraw.

In the wild places of this country, plants of the Madder Tribe abound. The Galiums, of which there are many species, occur on banks, heaths, and even walls, and are among the most common of plants. One of them, Galium verum, with loose bunches of pretty yellow flowers, is, in some counties, called *Cheese-rennet*, because of its having been formerly employed to curdle milk. *Woodruff*, a native rival in fragrance to the Heliotrope, is the Asperula odorata; it is found occasionally in woods, and is known by the long tube to its corolla, and the four small deciduous teeth of its calyx; otherwise it is very nearly a Galium. *Field Madder* (Sherardia arvensis) has a little purple blossom, and its fruit is terminated by the four permanent teeth of its calyx; it is a common annual in dry fields.

I need not ask if you have forgotten Compound flowers (Vol. I. *p.* 199. *Plate* XVII. 1.), for they are of such common occurrence, that to have had them once pointed out is to know them for ever. But I may ask if you recollect exactly what their structure is, because there are plants very like them at first sight, and which you must know how to distinguish. For example, Astrantia, which is one of the Umbelliferous Tribe, and Gilia, belonging to the Greek Valerians, have their flowers in heads, and might be taken for Compound flowers by an incautious observer. They are not, however, so likely to deceive you as the plants of the Scabious Tribe to which I have once already casually referred (Vol. I. *p.* 208.), and of which it is now time to speak more particularly.

Purple, or *Sweet Scabious* (Scabiosa atropurpurea),

is one of the most beautiful of our annual exotics, with its intensely deep purple corollas, and *Starry Scabious* (S. stellata), although not very pretty when in flower, is often found in gardens, where it is cultivated for the sake of its curious heads of seed-vessels, terminated by dry starry cups. In this country, the meadows, pastures, and corn-fields, are often enamelled by one or two native species, of which we will select what is commonly called the *Devil's-bit Scabious* (Scabiosa succisa, *Plate* XLIV. 2.); a strange name, which originated in a popular belief, in former days, that his Satanic majesty bites off the end of its roots; in proof of the truth of which, their black colour and abruptly-broken extremities are safely appealed to.

The root-leaves of this plant are obovate and undivided, those of the stem are coarsely toothed, while the uppermost are narrow, sharp-pointed at each end, and quite destitute of toothing. The flowers are a bright clear blue, and collected into round balls, at the top of long, slender, bristly peduncles (*fig.* 1.); in general appearance they very much resemble those of a compound flower, only they have not a distinct involucre; in the exact details of structure there are, however, several important differences, as you will see. To understand the matter fully, take one single floweret away from the others, and study it by itself; the remainder are like it. In the first place, you will remark, that it is subtended by a narrow sharp-pointed bract (*fig.* 2. *a.*), fringed with long delicate hairs. It appears to have a double calyx; the exterior being an inferior, pale, greenish-white cup (*fig.* 2. *b.*), with five angles and five shallow teeth; the interior (*fig.* 2. *c.* and *fig.* 4.) being a superior greenish disk, expanded into five purple hairy horns; of these two coverings the first is a little involucre, of the same nature as that in the Mallow (*Plate* VI. 1. *fig.* 3. *a.*), the second is the true calyx. The corolla is funnel-shaped, with its border divided into four nearly equal lobes (*fig.* 2); there are four stamens, which spread away from each other, without at all adhering (*fig.* 3.) either by the filaments or anthers. The ovary is one-celled, and what we technically call inferior; but it is a most unusual and instructive illustration of the correctness of the opinions of modern Botanists as to the real nature of a superior calyx (Vol. I. *p.* 28.). In the plant before us the ovary is a thin membranous case (*fig.* 5. *b.*), surrounded by the sides of the calyx, which, however, does not adhere to it, except quite at the orifice of its tube; but there the union is so complete that no trace of the separation lower down can be seen, except upon dissection. The style is a slender thread, curved upwards, and bearing a purple, narrow, hammer-headed stigma (*fig.* 4. *a.*). The ovule hangs pendulous from the top of the ovary (*fig.* 5. *a.*). The ripe seed-vessel is an oval seed-like body, terminated by five stiff, brown, hairy horns, and containing a pendulous seed, the embryo of which lies in albumen, with its radicle pointing to the apex of the seed-vessel (*fig.* 7.).

These details shew you that, notwithstanding the general resemblance of the Devil's-bit to a composite flower, it differs in having distinct stamens, and a pendulous seed, exclusively of all other circumstances.

This is, in reality, the difference between the *Tribe of Composite Flowers*, and the *Scabious* Tribe.

The most remarkable plant of this natural Order is the *Teasel* (Dipsacus Fullonum), the bracts of which are hard and sharp, and project beyond the flowerets, rendering the flower-head a cone of formidable spines. These heads are used in vast numbers in the carding of woollen cloths, and are found superior for that purpose to any artificial substitute yet invented.

EXPLANATION OF PLATE XLIV.

I. THE MADDER TRIBE.—*Cleavers, Goose-grass,* or *Whiptongue* (Galium Aparine).—1. A magnified flower.—2. A section of the same; *a* the green epigynous disk.—3. A stamen.—4. The ripe fruit.—5. A section of one half of a ripe fruit, shewing the embryo lying in the hard horny albumen.—6. An embryo separated, and inverted.

II. THE SCABIOUS TRIBE.—1. The *Devil's-bit Scabious* (Scabiosa succisa)—2. A floweret with the bract *a,* and the small involucre or involucel *b*, beyond which the calyx *c* is seen projecting.—3. A corolla cut open.—4. A pistil with the superior calyx; *a* the stigma.—5. A vertical section of the calyx, shewing that the ovary *b* does not adhere to its sides, except at the point; *a* the ovule.—6. The anther and upper part of a filament.—7. A section of a ripe fruit, surmounted by the calyx, and shewing the pendulous embryo lying in the midst of albumen.

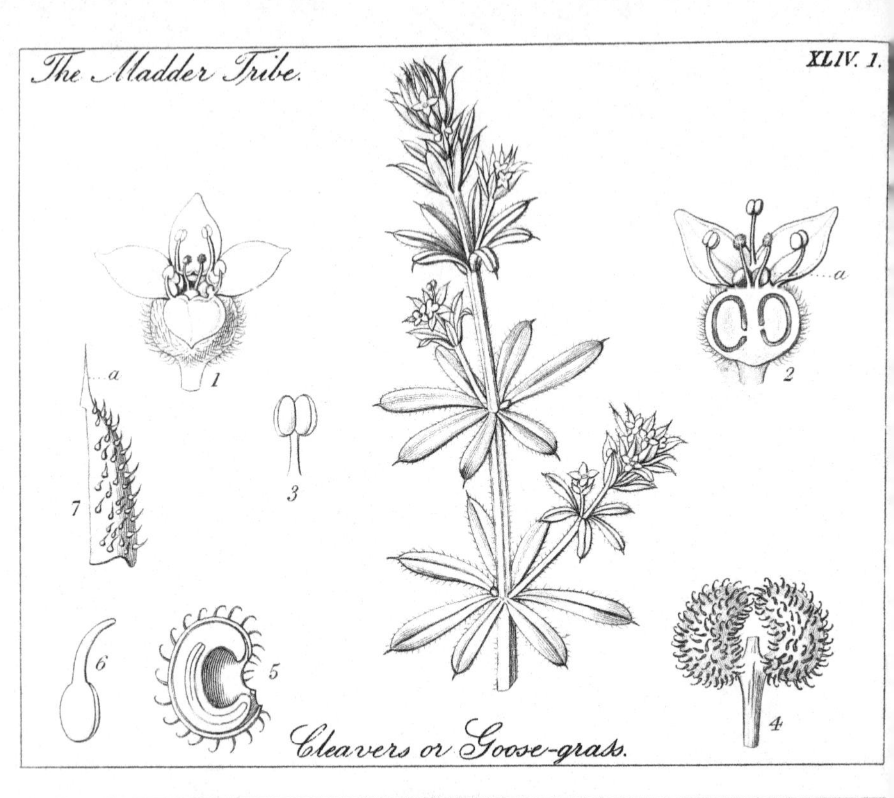
The Madder Tribe.
XLIV. 1.
a
1
2
3
4
5
6
7
Cleavers or Goose-grass.

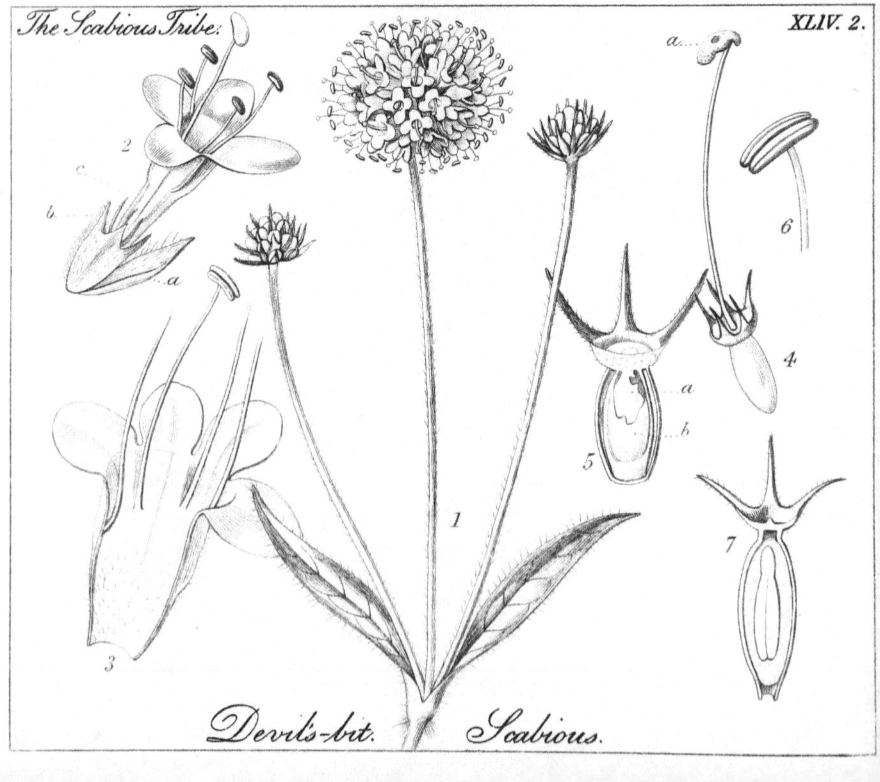
The Scabious Tribe.
XLIV. 2.
a
b
c
1
2
3
4
5
6
7
Devil's-bit. Scabious.

LETTER XLIV.

THE JASMINE TRIBE—THE ASCLEPIAS TRIBE.

Plate XLV.

I FORMERLY said something to you concerning the difference between the *Olive* and the *Jasmine* Tribes (Vol. I. *p.* 168.); and perhaps the brief remarks then made upon the method of distinguishing them may have satisfied you. Nevertheless let us not pass the Jasmine by with inattention, for surely so lovely a plant deserves something more than a careless glance of recognition.

The *White Jasmine* (Jasminum officinale), the pride of the cottager, and the envy of the citizen, within whose smoky streets no art can make it flourish, is a native of the mountains of India, whence years ago it found its way to the Persians and Arabs, who called it Yasmeen, and thence passed to Europe. Its leaves offer a good example of what we call *unequally pinnated*, or *pinnated with an odd one;* that is to say, they consist of several pairs of leaflets (*Plate* XLV. 1.), with an odd leaflet at the end. The leaves are opposite each other on the stem, and have no stipules. The virgin-white odoriferous flowers grow in little sessile clusters, or umbels, at the end of short branchlets. The calyx is inferior (*fig.* 3.), divided into five narrow awl-shaped segments, and covered externally with glandular down. The corolla is salver-shaped, with a long yellowish tube, and a border divided into five sharp-pointed lobes, which do not fit to each other by their very edges as in the Olive Tribe, but overlie each other, and are twisted together in the bud state (*fig.* 1. *a.*); or, in fewer words, the æstivation is imbricate, and not valvate.

There are only two stamens, arising from near the middle of the tube of the corolla (*fig.* 2.). This is a degree of irregularity much beyond that of the Olive Tribe, which has only four segments of the corolla, and is more analogous to what occurs in the Foxglove and neighbouring Tribes. In many species of Jasmine it is carried further still; for the corolla has sometimes six, seven, eight, or even twelve divisions, and it must therefore be considered to have a tendency to form two or even three rows of petals; or else to develope a part of its stamens in the form of petals, and even to produce a second row of true petals in addition. Thus the scheme of organization in common Jasmine will be—

S S S S S
P P P P P
. . s s .

in a Jasmine with seven segments to its corolla—

S S S S S
P P P P P
. *s* s s *s*

and with twelve segments—

S S S S S
P P P P P
P P P P P
s s s *s*

so that those species only can be considered complete, in the number of their parts, whose corollas consist of eight or thirteen segments.

The ovary of the Jasmine (*fig.* 3.) is superior, and contains two cells, with an ascending ovule in each; another mark of distinction from the Olive Tribe, in which the ovules are pendulous (*Plate* XIII. 2. *fig.* 5.). The style is erect and slender; the stigma a fleshy, glandular, two-lobed club. The fruit (*Plate* XLV. 1. *fig.* 4.) is a black oblong berry, containing one perfect and one abortive seed (*fig.* 5.); the embryo is covered over by the seed-coat without the aid of any albumen.

Such is the common sweet white Jasmine, and such, in all essential points, is the remainder of this fragrant genus. The species differ in respect, 1. to the manner of growth, some climbing, and others forming mere bushes; 2. to their leaves, some of which are undivided, and their form various in various species; 3. to the colour of their flowers, which, although usually pure white, is sometimes yellow; and 4. to their corolla, the number of whose divisions is, as has lately been mentioned, extremely variable. All of them, however, have monopetalous corollas, with several *imbricated* segments, only *two* stamens, and a superior, succulent fruit, with one or two erect seeds.

The Jasmine Tribe consists of few plants besides Jasmines themselves. The most remarkable is the *Tree of Mourning* (Nyctanthes Arbor Tristis), or *Hursinghar*, an Indian tree of small size, whose "exquisitely fragrant flowers, partaking of the smell of fresh honey" (I quote Dr. Roxburgh's words), open at the close of day, and fall off before sunrise, strewing the ground with their remains, and furnishing to the Indian girls the materials with which they decorate their hair. After the flowers have passed away, this tree becomes ragged and shabby, assuming a melancholy appearance, as if in grief for the loss of the fragrant treasures that it once dispensed with so lavish a hand. This circumstance, and the dark hours of night which the plant selects for displaying its charms, have doubtless given it the name of Arbor tristis, or the tree of mourning. It is known botanically from a Jasmine, by its fruit being a dry seed-vessel, instead of a succulent berry.

You may well be puzzled with the plant enclosed in your letter of yesterday; and you are right in your conjecture that it is not even alluded to, in any part of our previous correspondence. It is the *Pink Asclepias* (Asclepias incarnata), and forms the type of the Natural Order of that name. Its flowers are most curiously constructed, and may well embarrass you even to name the parts of which they consist. After you have received this letter, gather a fresh cluster of the blossoms, and follow me in the ensuing description.

In gathering it, you will find milk flow abundantly from the wound; in this plant the milk is white, but in one species inhabiting the woods of Sierra Leone, it is of the colour of blood. If that plant had but grown in Palestine, it might be supposed to represent the enchanted tree, which so surprised Tancred in the sorcerer's wood—

> " When, dreadful to his view !
> The wounded bark a sanguine current shed,
> And stain'd the grassy turf with streaming red."

This milky blood, whether white, red, or any other colour, abounds in the substance called Caoutchouc or Indian Rubber, a large proportion of which is actually procured from plants botanically related to the Asclepias.

You will readily distinguish the calyx, which consists of six, narrow, hairy sepals, spreading back from the corolla (*Plate* XLV. 2. *fig.* 5. *a.*). Their purple colour betrays the petals (*fig.* 2. *a.*), which spread widely away from the centre, adhering at the base only, into a short tube, and therefore constituting a monopetalous corolla. From the middle of the tube there rises a pentagonal column (*fig.* 2. *b.*), forming the base of five other concave petals (*fig.* 2. *c.*), which stand erect, and collect into a sort of pink coronet (*corona*) to the flower; from the inside of each of these coronet-petals, springs a firm, solid horn, curving forwards towards the centre (*fig.* 3. & 4.). The monopetalous corolla is therefore composed of two whorls of petals, of which the outer are flat and spreading, the inner concave, erect, and horned internally.

Cut away the true petals and those of the coronet; you will find that the pentagonal column consists of five purple-green anthers (*fig.* 5. *c.*), having no filaments, where they come in contact projecting into five whitish angles, adhering firmly by their faces to a pentagonal, flattish, fleshy, red and green table, which they surround (*fig.* 7. *a.*), and having each a whitish, membranous termination, which curves over the table aforesaid (*fig.* 5. *e.*). In the next place, carefully remove two of the anthers, turning them on their backs (*fig.* 6.); you will find that each is two-celled, and that the pollen of the contiguous cells of two different anthers, forms two orange-coloured bags (*fig.* 6. *a.* and *fig.* 9.), which are very loose in their cells, and adhere to a blackish, oval gland, that belongs to the angle of the table aforesaid (*fig.* 7. *b.* 6. *b.* & 9. *b.*); so that when you open the anthers, you see the bags dangling from the gland like a pair of yellow pouches (*fig.* 9.).

After all this apparatus is removed (as at *fig.* 7.), you have a view of the pistil, consisting of two ovaries placed in close contact, and each containing a large, fleshy placenta, covered with ovules (*fig.* 8.). To each ovary is a single style, which is placed parallel and in contact with that of its neighbour, without uniting to it (*fig.* 7. *c.*). The styles are held together by the fleshy five-cornered table that surmounts them (*fig.* 7. *a.*), and which stands in the place of a stigma, without exactly being one; for the influence of the pollen is not communicated to the ovules through its tissue, as in true stigmas, but somewhere about the point where the style and the table join (*fig.* 7. *d.*).

When the corolla and stamens have fallen off, the table and styles give way, the two ovaries diverge, and if both of them continue to grow, you will find, when the seed-vessel is ripe, that it consists of two, dry, tough cases, opening by the face, and placed almost at right angles with each other, so that the seed-vessels and stalk together, form the figure T. But it often happens that one ovary shrivels up and disappears; in that case, the other grows upright upon its stalk (*fig.* 10.), as is usual in other plants. In the inside of the seed-vessel is a large number of flat, brown seeds, terminated by a delicate silk tuft (*fig.* 10.), and containing a thin, flat embryo, without any albumen (*fig.* 11.).

It must be quite plain to you, that at least three circumstances will separately characterise the Asclepias Tribe; for no other Monopetalous order has either, 1. the pollen adhering into bags, or, 2. the anthers adhering firmly to a stigma-like table, or, 3. the corolla augmented by a coronet, or second row of petals.

You must not, however, expect that all the tribe will agree in the nature of their coronet; some have only a single row of secondary petals, as the plant now before you, others have two or even three rows, in various states of combination or developement. For the purpose of studying these matters, you should examine the curious speckled flowers of the *Stapelia*, and the honey-dropping, waxen blossoms of the *Hoya*. The former, indeed, will generally repel you by their intolerable smell, if you wait till they are naturally expanded; but if you cut them open some days previously, you can examine them without inconvenience.

Cynanchum and *Periploca*, are other common genera

of the same tribe, which you will easily procure for study.

Very nearly allied to the Asclepias Tribe, are the poisonous *Apocynums*, represented in the gardens by the *Periwinkle* (Vinca), the *Oleander* (Nerium), and the Apocynum itself. They agree with the Asclepias Tribe in their milky juice, and their appearance, but differ in having the stamens free from each other and from the stigma, the pollen in its usual state, instead of being collected into bags, and in the want of any coronet of secondary petals, except now and then a single row of scales, growing in the mouth of the tube of the corolla.

As the plants of the Apocynum Tribe, with the exception of those now mentioned, are not likely to fall in your way, it is not necessary for you to be detained with any account of them.

EXPLANATION OF PLATE XLV.

I. The Jasmine Tribe.—1. *Common white Jasmine* (Jasminum officinale), *a*, a flower-bud, shewing the imbricated twisted æstivation. —2. A corolla magnified, and opened to shew the position of the stamens.—3. A section of the calyx and ovary, exhibiting the position of the ovules.—4. A ripe fruit.—5. The same cut across.

II. The Asclepias Tribe.—1. The *Pink Asclepias* (Asclepias incarnata).—2. A flower magnified; *a a* petals, *b* the column that supports the coronet of secondary petals, *c c*.—3. One of the petals of the coronet cut off the column.—4. A section of the same, to shew the origin of the horn.—5. A flower from which all the petals have been removed; *a a* sepals, *b* the column of the coronet, *c c* anthers, *d d* the projecting angles formed by the sides of the anthers, *e* the membranous appendages at the tip of the anthers, *f* the glands to which the pollen-masses adhere.—6. Two anthers turned on their backs; *a a* a pair of pollen-masses, *b* the glands to which they adhere.—7. A pistil; *a* the stigma-like, pentagonal, fleshy table, to which the anthers adhere, *b b* the places where the pollen-glands adhered, *c* the parallel styles, *d* the place through which fertilization is conveyed from the pollen to the ovules.—8. A transverse section of the two ovaries.—9. A pair of pollen-masses, adhering to the gland *b*.—10. Ripe seed-vessels.—11. A seed cut across, to shew the embryo.—12. A perfect seed, with the feathery tuft that terminates it.

LETTER XLV.

THE BIRTHWORT TRIBE—THE ARUM TRIBE.

(Plate XLVI.)

Did you ever remark a broad, round-leaved, twining plant, near the entrance to the flower garden, on the right hand, next the little rock-work for Sedums, with dingy, brownish, lead-coloured flowers, bent almost double in the middle, and only to be discovered by a careful search among the leaves? It is a plant called Aristolochia Sipho, or, in English, the *Siphon-flowered Birthwort*, and belongs to the same natural order as the *Wild American Ginger* (Asarum canadense), that little, round-leaved, stemless plant, which forms two or three clusters among the Azaleas, in front of the library window, and whose cup-shaped brown flowers I remember shewing you, as carefully hidden among the leaves as if they had been, what they really look like, the nests of some fairy bird. These plants are all of them excessively curious, because of the strange form of their flowers; most of which are singularly mottled or veined with brown or purple, and some of which are quite gigantic in their dimensions. Humboldt saw the children of an Indian village, wearing the helmet-shaped flowers of one

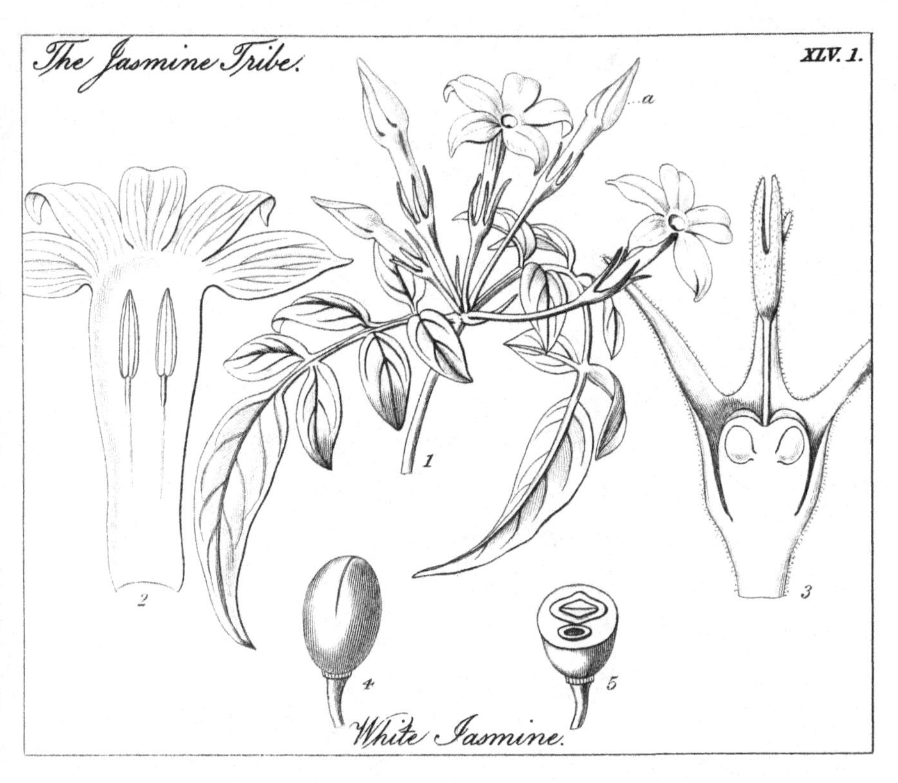
The Jasmine Tribe.
XLV. 1.
a
1
2
3
4
5
White Jasmine.

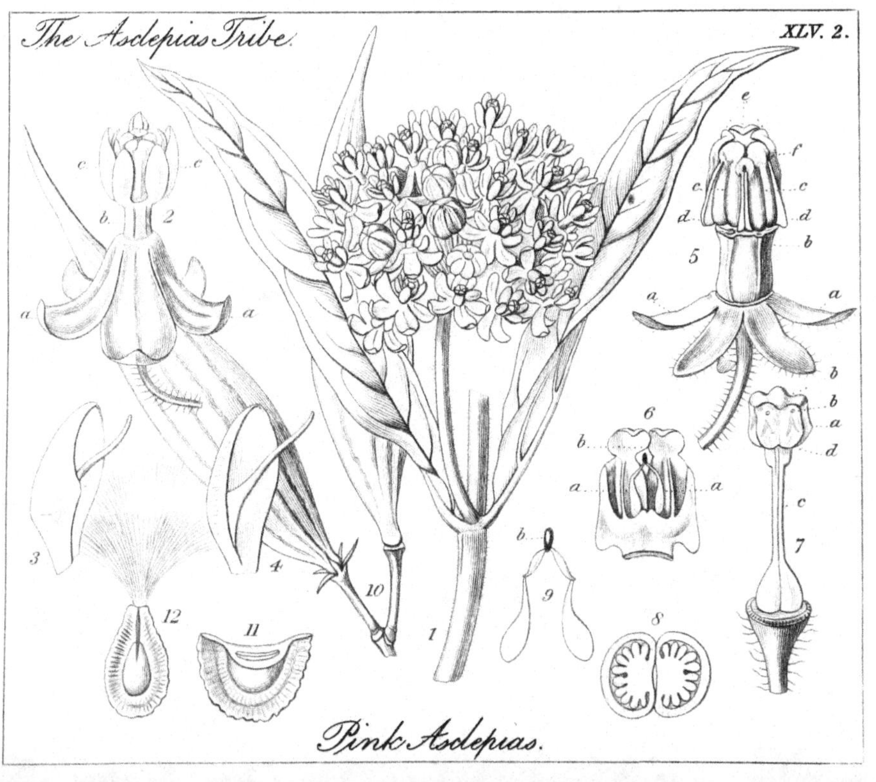
The Asclepias Tribe.
XLV. 2.
c
c
b.
2
a
a
e
f
c.
c
d.
d
b
5
a
a
b
b
a
d
6
b.
a
a
c
7
b.
9
8
3
4
10
12
11
1
Pink Asclepias.

species, instead of hats, and Brazilian kinds have been discovered of scarcely a smaller size.

These plants are brought to my recollection, by a quantity of blossoms of Aristolochia trilobata, the *long-tailed Birthwort,* that some unknown friend has just sent me from her hothouse. With this letter you will receive a portion of them, which we will proceed to examine systematically (*Plate* XLVI. 1.). We will take no notice of its twining stem, nor of its leaves, for these organs vary so much in different species, as to form no part of the distinctive characters of the order, but we will confine ourselves to the fructification.

The flower is a long tubular calyx, strongly veined and ribbed, curved back in the middle, so much as almost to be bent double, pale, livid, brownish-yellow externally, and deep chocolate brown in the inside, and at the upper end (*fig.* 1.). At the lower end it is inflated; and at the very base it is extended into six little horns or spurs (*fig.* 1. *c.* & *fig.* 2. *b.*). At the upper end it is very much dilated and puckered; on one side (*fig.* 1. *e.*) it is deeply notched; on the opposite side it is extended into a flat, twisted strap, thirteen or fourteen inches long, which, when the flowers are on their branches, hangs down like the tail of some animal; one might even fancy it belonged to a mouse, whose body was secreted in the cup of the flower. This curious calyx is quite at variance with any thing, however irregular in structure it may be, that we meet with in the same part, in the rest of the vegetable kingdom. Botanists seemed generally agreed in considering it composed of three consolidated sepals, of which two are rounded and one only produced into a long appendage or tail. This opinion is founded partly upon the prevalence of the number three in the other organs of fructification, partly upon the regular flowered genera of the Birthwort Tribe having a calyx of three divisions, and in some measure upon the theory, that a calyx is in all cases to be considered a whorl of sepals. It may, however, be fairly doubted whether in the genus Aristolochia, the calyx is really formed of more than a single sepal, or leaf, rolled together into a tube, and, in the present species, extended at its point into a tail. But to this I shall advert again.

At the bottom of the cup of the calyx stands a short, club-shaped column (*fig.* 2.), split into six lobes at its point; and consisting of six anthers, adhering to a style and six-rayed stigma which they conceal. Each anther (*fig.* 3.) is a fleshy, somewhat shrivelled, sharp-pointed connective, on the outside of which are planted two parallel cells, which consequently are turned away from the stigma, and face the inside of the calyx. The ovary is placed beneath the calyx, in the form of a club-shaped, twisted stalk (*fig.* 1. *d.*); it contains six cells (*fig.* 4.), in each of which is a long row of ovules, attached obliquely to the placenta. With the seed-vessel of this species I am unacquainted; but in others it is a large pear-shaped capsule, opening by six sutures at the sides, and allowing the seeds to escape through a sort of coarse grating, produced by a laceration of the dissepiments. The seeds are thin, flat, and dark brown (*fig.* 5.), and contain a small, dicotyledonous embryo, at the base of hard, horny albumen (*fig.* 6. & 7.).

Asarum, the only other genus of this order you are likely to meet with, has a regular three-lobed calyx, and its stamens are distinct from each other; the adhesion of the stamens into a central column, does not therefore form any part of the essentials of the Birthwort Tribe, which is characterised by its inferior, six-celled fruit, its six stamens, and by its tubular calyx without corolla, divided into either one or three lobes; so that the type of its tructure is essentially ternary, or thus,

S S S . S .
. . . . . .
s s s or s s s
s s s s s s
c c c c c c
c c c c c c

which among Dicotyledons is very uncommon.

It is, as you know, chiefly in Monocotyledons or Endogens, that the number three prevails in the parts of fructification, and it is not a little curious, that the stem of Aristolochias should be almost intermediate in structure between that of Exogens and Endogens. It has the medullary processes of the former, and consequently their pith; but it wants the concentric layers in the wood, which is formed of bundles of woody matter, collected indeed into wedges, but plunged down into a pithy substance, as in Endogens. The Birthwort Tribe may therefore be considered one of several cases, where the structure peculiar to one class assimilates itself to that of the other.

A case of this sort, where Aristolochias themselves may be considered as typified among Endogens, occurs in the Arum Tribe (*Plate* XLVI. 2.). You probably know this tribe already, from the common *spotted Arum* (A. maculatum) of our hedges, or the *speckled Dragon Arum* (A. Dracunculus) of the gardens. These two species, at least, are so very common, that if you do not yet know them, you can have no difficulty in procuring them for examination.

The *Arum Tribe* consists of stemless or long-stemmed plants, whose internal structure is strictly that of Endogens, but whose leaves bear more resemblance to those of Exogens; it is, however, to be observed, that the lobed figure of the leaves, and their branched veining, to which the resemblance is due, need not be confounded with the netted veining of Exogens, because in Arum, the veins are branched rather than netted, and are in a great measure destitute of the lateral, minute branchlets, to which the peculiar appearance of Exogenous leaves is chiefly owing. Many of these have large, tuberous, under-ground stems, which, although acrid, and even poisonous when raw, nevertheless, by slicing, washing, and cooking, become fit for food, and are actually so employed, in England only in a few places, or in times of scarcity, but in tropical countries, as a common, every-day, esculent vegetable. Their foliage is generally more or less lobed, and sometimes very curiously, but is so much diversified, that it can hardly be said to offer any certain mark of recognition. The great and striking feature of the natural order resides in the *spathe* and *spadix*. As these terms are new to you, they must be explained before we proceed further.

A *spathe* is a leaf, usually coloured, but sometimes green, which is rolled up round a spike of flowers; it is, in fact, a sort of large bract.

A *spadix* is a fleshy spike, covered all over with flowers, and enclosed in a spathe.

In all Araceous plants, the flowers are collected upon a spadix, and are enclosed in a spathe. Both these parts, in particular species, have most extraordinary appearances. The spathe, for example, is sometimes a foot and more in diameter, forming a huge vegetable bell, of which the spadix would be the clapper, if the spathe were not erect; it is often stained with the deepest and richest colours; and in some cases it is extended on one side into a long slender tail, very much like that of the calyx in the long-tailed Birthwort. The spadix, on the other hand, is either covered all over with flowers, in which case it makes no unusual appearance, or it is naked at the point and then assumes the strangest shapes, which sometimes, moreover, glow with all the colours of the spathe. Thus in the Dragon-Arum it is a long purple horn, standing up, and projecting from a large, deep-purple spathe; in others it hangs down from the spathe like a slender tail; and in some cases it is enlarged into a disgusting, fungus-like, livid excrescence.

The *common spotted Arum* (Arum maculatum), will give you a sufficiently correct idea of the structure of the Arum Tribe. It has a smooth, erect, oblong spathe (*fig.* 1.), green outside, whitish inside, and unrolling to expose the point of the spadix (*fig.* 1. *a.*), which children call the lady riding in her coach. If you extract the spadix, you will find it a long, soft, fleshy branch, the upper part of which is quite naked, and the lower part covered with naked flowers. At the bottom (*fig.* 2. *b.*) stand several tiers of round ovaries; above them are placed two or three rows of abortive ovaries, in the form of horned, pear-shaped bodies (*fig.* 2. *c.*); then appears a crowd of stamens (*fig.* 2. *d.*); and above those is again collected a small cluster of abortive ovaries (*fig.* 2. *e.*). The ovaries are so many naked fertile flowers, the stamens are each a naked sterile flower; and the inflorescence is, in strict technical language, a crowded monœcious spike, wrapped up in a large leafy bract.

The ovary is puckered and hollowed out at the apex, for a stigma (*fig.* 3.), and contains two ovules growing from the side of a single cell (*fig.* 4.). The stamen has a short thick filament, with two round lobes, placed obliquely on its end, for an anther (*fig.* 5.).

The fruit ripens in the form of a spike of orange-coloured, roundish berries (*fig.* 6.), each of which contains a single seed (*fig.* 7.), enclosing a monocotyledonous embryo (*fig.* 8.), surrounded by farinaceous albumen. On one side of the embryo is a narrow slit (*fig.* 8. *a.*), at the bottom of which lies the minute point (*fig*. 9. *a*.), or plumule, which eventually becomes the new stem.

Such is the structure of the spotted Arum. The other genera differ in the spadix being altogether covered with flowers, or in the absence of abortive ovaries, or in the internal structure of the anther and ovary, or even in that of the style and stigma; but the spathaceous inflorescence distinctly marks the order in all cases.

EXPLANATION OF PLATE XLVI.

I. The Birthwort Tribe.—1. Leaf and flower of the *long-tailed Birthwort* (Aristolochia trilobata), natural size; *a* the stipules; *b* the strap-shaped tail of the calyx; *c* the horns at the base of the calyx; *d* the ovary; *e* the notch on one side of the border of the calyx.—2. The column of stamens *d*, seated in the base of the calyx *a*; *b* the horns of the calyx; *c* the ovary.—3. An anther separated from the other five. —4. A transverse section of the ovary.—5. A seed of another species of Birthwort, natural size.—6. The same magnified, with half the skin cut off, to shew the embryo *a*, lying in horny albumen.—7. The embryo.

II. The Arum Tribe.—1. Spathe of *Spotted Arum* (Arum maculatum), natural size, with the point of the spadix at *a*.—2. A spadix taken out of the spathe; *a* the base of the spathe; *b* ovaries; *c* abortive ovaries; *d* stamens; *e* abortive ovaries; *f* part of the stalk of the naked head of the spadix.—3. An ovary.—4. A transverse section of the same. —5. A stamen.—6. The spadix covered with ripe fruit; *a* the withered remains of the spathe.—7. A ripe seed.—8. A longitudinal section of the same, shewing the embryo lying in albumen; *a* the slit communicating with the plumule.—9. The monocotyledonous embryo, with a portion of its root-end sliced away, and shewing the conical plumule *a*.

LETTER XLVI.

PITCHER-PLANTS—VEGETABLE ANATOMY.

Plates XLVII & XLVIII.

I MUST not dismiss the Birthwort Tribe without adverting to those curious vegetables called *Pitcher-plants*, in the East Indies, and to the *Monkey-cups* they bear. The production of hollow bags instead of leaves, is not a very uncommon occurrence in plants; in *Dionæa* a preparation is made for their formation by the dilatation of the leaf-stalk; in *Side-saddle flowers* (Sarracenias), the edges of the petiole are rolled up and united into a cup, over which the end of the leaf curves, as if to cover it; in some plants the bracts are changed into bags which hang down amongst the flowers; and in an East Indian plant called Dischidia Rafflesiana, which climbs to the top of the highest trees in the forests of Penang, the upper branches are loaded with clusters of tough, fleshy, leathery bottles, filled with water, into which roots, protruded from the branches, dip their points to drink.

Not only is *Sarracenia* found in our gardens now and then, and *Cephalotus*, a New Holland plant, whose singular pitchers are beautifully fringed and veined; but Nepenthes itself, the true *Pitcher-plant* of the

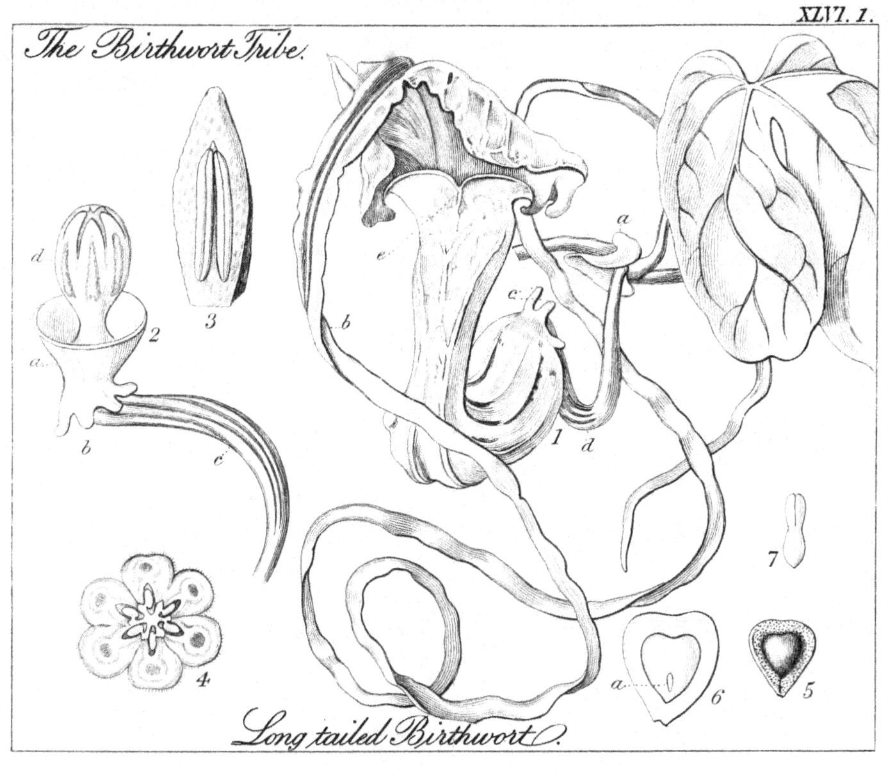
XLVI. 1.
The Birthwort Tribe.
Long tailed Birthwort.

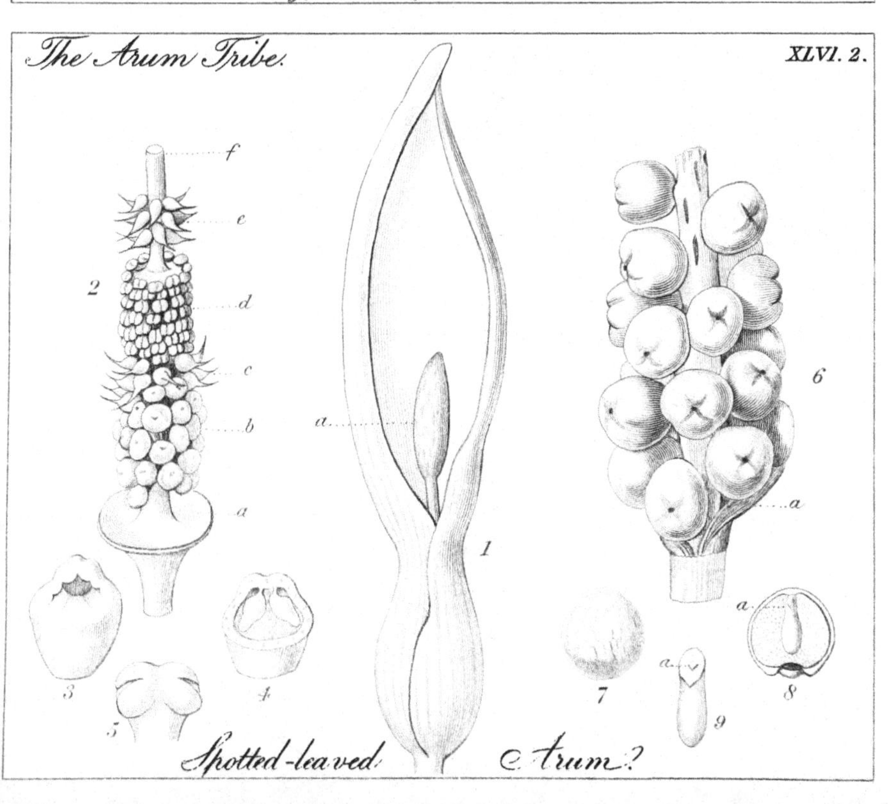
The Arum Tribe.
XLVI. 2.
Spotted-leaved Arum.

East Indies, is successfully cultivated in several places. A large branch which I have received from Wentworth, through the kindness of Lord Fitzwilliam, gives me an opportunity of describing it to you in detail; and I know no plant that better deserves to be understood.

The stem of Nepenthes distillatoria, the only species as yet in the possession of cultivators, forms a slender woody stem, growing ten or twelve feet high, or probably much longer, and supporting itself upon surrounding plants, by means of its numerous tendrils. In its native country it inhabits swampy situations, and consequently, in a hothouse, it must be treated accordingly. At the lower part of the stem there appear a few leaves of a bright green colour, a little curved back at the point, where they are rounded off, tapering to the base and half embracing the stem; these leaves vary in length from one to four or five inches, and have nothing in their appearance to distinguish them from ordinary leaves. But higher up the stem, the leaves grow much longer, and taper into a tendril at the point, from which is suspended a long funnel-shaped, green cup, often as large as a three ounce vial, covered by a lid, and sometimes containing water. At first it is entirely closed up by the lid, but after a time the latter separates, except by its hinge, and merely overhangs the mouth of the cup, which is bordered by an exquisitely beautiful, stiff, crimped frill, which curves inwards, and forms a broad ledge on which the sides of the lid may rest. In all cases the pitchers contain fluid at some time or other; but after they are once opened it usually dries up; so that the tales which are current about their being sought by wild animals, especially monkeys, for the water they contain, must be received with some suspicion. The use of the water is altogether unknown, nor indeed are Botanists generally aware by what apparatus it is secreted. There is, however, a peculiar glandular structure in the inside of the pitchers, which is the more probably connected with the secretion, as it is not found on any other part of the Nepenthes, nor, so far as I know, in any other plants.

If you observe attentively the inside of the lid of the common garden species, or peep down into the pitchers, you will find the surface distinctly marked by inequalities, which give it somewhat the appearance of shagreen. Placed under a microscope, the inequalities prove to be caused by the presence of an infinite number of oval, dark brown glands (*Plate* XLVII. *fig. C. a.*), lying in the midst of the fine, compact, cellular substance of the cuticle. The cells of the latter are tolerably regular, lozenge-shaped hexagons, except at the edges of the glands, where they become perceptibly smaller and rounder (*fig. C. b.*); and, what is very remarkable, the cuticle, instead of spreading over the glands, leaves them quite naked, so that, when it is stripped off the leaf, it is riddled with regular oval holes (*fig. C. b. b. b.*) corresponding with the glands. If, instead of examining merely the surface of the interior of the pitcher, you make a section of it, perpendicular to the surface, and through one of the glands (*fig. D.*), it will then be seen that the gland (*a*) is really an oblong kernel, of hard, brown, minute cells, lying upon a quantity of thin-sided vesicles of the parenchyma, and kept in its place by the edge of the tough cuticle, which projects a little over the edge, and holds it firmly down; there is the more necessity for this arrangement, in consequence of the gland having no connection with the tissue it lies upon, further than it gains by being in contact with it. As glands are so often secreting organs, is it not probable that the secretion of fluid inside the pitcher of Nepenthes, may be owing to their presence? I have stated, that be their office what it may, they never occur any where except on the inside the pitcher; in Nepenthes distillatoria, they are not found near the top, although they are abundant on the inside the lid; in other species, the lid seems to be quite free from them, while the whole of the interior of the pitcher is covered with them. I have also, in one solitary instance, seen three of them on the outside of a pitcher near its base.

It is not merely in the cuticle of its pitchers, that Nepenthes has a curious anatomy. It is extremely well worth examination in other parts, and as we have all our microscopical apparatus in readiness, we may as well continue the investigation. Let us begin by making a very thin, transverse slice of the stem; this will shew you, that whether it is the soft parenchyma of a leaf or the firmer tissue of the bark, or the delicate and filmy cuticle, or the solid wood itself, all the parts of a plant consist of cells and tubes variously arranged. Having placed your slice on the table of the microscope, in water, and illuminated it by light thrown from below, first remark the structure of the bark ; it is a thick, firm layer of hexagonal cells, part of which (A. *f.*—*g.*) are arranged in one way, and part (A. *e.*—*f.*) in another, so that a strip of the bark might without much difficulty, be split into two plates. Among the green cells of the bark, you will remark a few round white points : these are the mouths of fine, spiral-coated tubes, or *spiral vessels*. Between the bark and the wood is a thick layer (A. *d.*—*e.*) of exceedingly delicate, thin, green cells, in which you may discern the round mouths of other tubes of various sizes ; these are other spiral vessels of very large size, and in such abundance, that they look like a stratum of tow, between the wood and bark ; each of these large spiral vessels is formed of four threads, twisted spirally. Next the spiral structure comes the wood, the outside of which (A. *c.*—*d.*) is hard, compact, and homogeneous, and then becomes, towards the centre (A. *b.*—*c.*), more open, with a quantity of unequal, round, or oval perforations, which are also the mouths of large spiral vessels ; finally, you come to the pith (A. *a.*—*b.*), consisting of thin-sided, large cells, in which are more mouths of vessels. All this is highly curious, and shews you what an infinite multitude of forces, represented by these little organs, are required to maintain the life of Nepenthes.

You will not, however, form a correct notion of their real nature, unless you also examine a longitudinal slice of the same part of the stem (*Plate* XLVII. B.) ; hitherto you have only seen the ends of the cells and tubes; you are next to observe their sides; or otherwise you will not distinguish between tubes and cells. To begin again with the bark. You will now find that the cause of the different appearances in the two layers of the bark is owing to the outer layer (B. *f.*—*g.*) consisting of round cells, while the inner consists of long cells (B. *e.*—*f.*), whose principal diameter is parallel with the stem; of these two layers, the outer is purely parenchymatous, and analogous to the cortical integument, the inner is partly woody and analogous to the liber or inner bark of other Exogens. You will next see that the spiral stratum (B. *d.*—*e.*), is composed exclusively of thin roundish cells, and spiral vessels of the largest size; that the compact, homogeneous outside (B. *c.*—*d.*) is exclusively composed of woody tubes; that the wood itself (B. *b.*—*c.*) consists externally of woody tubes, which gradually, as they approach the pith, acquire an hexagonal form; and that in addition to the small spiral vessels lying amongst them, are some jointed, dotted tubes, which were not before distinguished; finally, that the pith is really composed of nothing but large, round polygons, mixed with great spiral vessels, as at first appeared.

You must not suppose that, because the Nepenthes is an Exogenous plant, therefore all other Exogens have exactly this structure. On the contrary, Nepenthes is one of the greatest anomalies I am acquainted with, and stands quite alone, so far as observation has yet gone, in several parts of its anatomy. For instance, no other known plant has spiral vessels any where except in the woody parts; Nepenthes produces them not only in the pith and the bark, but actually possesses a peculiar organ, as it would seem, expressly formed for their more abundant developement; namely, the cellular stratum between the wood and bark. My object, therefore, in bringing these points to your notice, is not so much to illustrate general structure, as to acquaint you with a great singularity of structure.

If you now proceed to examine the cuticle, you will find even there a circumstance which is very unusual. The stomates on the outside the pitchers, and on the upper side of the leaf, are quite different. The cuticle of the upper side of the leaf (*Plate* XLVIII. D.), consists of lengthened meshes formed by the union of long cells; and among them are placed colourless, oval stomates (D. *a.* and B. *a.*), formed of a pair of parallel cells, and containing a good many particles of semi-opaque matter. But on the outside of the pitchers, the stomates are different in form and colour; the cuticle of this part has rounded meshes (*Plate* XLVIII. C.), among which lie roundish reddish stomates (C. *b.* and A. *a.*), not appearing to contain glandular matter, and consisting of four cells, of which the two central ones are much deeper coloured than the others. Moreover, below each of these stomates, in the inside of the leaf, are arranged six or seven angular, deep-red cells, which form a sort of internal gland, resting upon the stomate (C. *b.*). This circumstance seems connected with the glandular structure of the inside of the pitcher, and possibly will be hereafter found another part of some wonderful adaptation of means to ends, which, although not capable of explanation in the present instance, we may feel perfectly persuaded of the existence of.

A slice of the firm tendril of this plant is so easily obtained, and shews so well the machinery by which that slender part bears its heavy pitcher, that I am sure you will be sorry to miss the opportunity of studying it. Take the finest imaginable transverse slice, and cut out of it a wedge (*Plate* XLVIII. E.), the top of which shall be the circumference, and the point the centre of the tendril. You will find that it is composed chiefly of roundish cells, the principal difference in which is, that those next the centre (*e.*—*f.*) have thinner and weaker sides than those next the circumference (*e.*—*d.*); and that the whole is bound together by a tough cuticle of small thick-sided cells (*d.*—*d.*). If the tendril were really composed of nothing more than this, it would have none of the requisite toughness and elasticity, either to support the weight of the plant, or to carry the pitcher; on the contrary, it would be brittle, like a piece of pith or a fungus. But upon looking more carefully at the section, you will perceive, near the centre, four or five little collections (E. *c.*) of thick-sided cells, surrounding a solid half moon (E. *f.*), and a small number of light, open, oval, or round spaces (E. *c.*), which you now know are the mouths of vessels; you will further note that the convexity of the half moon is towards the circumference of the tendril. A little way off the centre, towards the circumference, you will find from sixteen to twenty more of these appearances (E. *f.*). They are caused by your having cut through the ends of highly elastic cords, consisting of spiral vessels (*b.* & *c.*), strengthened by a quantity of woody fibre (*f.*), and surrounded on all sides by tough, thick-sided cells; respiration goes on through the spiral vessels, circulation through the woody tubes, which also give strength and elasticity to the cords, and digestion through the surrounding cells. Moreover, near the circumference of the tendril (E. *a.*) these cords are repeated on a smaller scale, spiral vessels being placed in the centre, thick-sided cells on the outside, and a few tough, woody tubes immediately in contact with the spiral vessels; the object of these is no doubt to strengthen the tendril still further, and to do away with all possibility of the cords near the centre being accidentally broken.

Thus you see nature provides not fewer than sixty or seventy cords or muscles, each of a most wonderful degree of completeness, to give its requisite strength to a tendril, the diameter of which does not exceed the twelfth part of an inch. I am sure you will now agree with me, that however admirable the contrivances are, which readily meet the eye in the vegetable kingdom, there is something still more wonderful in the hidden and microscopic machinery, by which their organs are set in action.

EXPLANATION OF PLATE XLVII.

Anatomy of Nepenthes.—A and B sections of a stem; *a—b* the pith containing spiral vessels, lying in cellular tissue; *b—c* wood, consisting of long, lozenge-shaped, thick-sided cells, passing into rounded cells as they near the pith, of small spiral vessels and dotted vessels* (or vasiform tissue) intermixed; *c—d* a homogeneous layer of woody tissue; *d—e* large, lax, thin-sided, cellular tissue, forming, with large spiral vessels, a layer between the wood and bark; *e—f* the liber or inner bark, and *f—g* the cortical integument, or outer bark, containing fine spiral vessels.——C. A portion of the cuticle of the inside of a pitcher, with the glands *a*, and the openings in the cuticle *b b*, left when the glands are removed.——D. A section of the pitcher, made perpendicular to the cuticle of the inside; *b b* cuticle closing in the gland, which is evidently a kernel of small hard brown cells.

EXPLANATION OF PLATE XLVIII.

Anatomy of Nepenthes, continued.——A. A highly magnified view of a stomate *a*, and a portion of the cuticle on the outside of a pitcher.—C. The same less magnified, seen from the under side; *a* the stomates; *b* the purple gland which reposes upon the stomates.—B. A highly magnified view of a stomate *a*, and a portion of the cuticle of the upper side of the true leaf.—D. The same less magnified; *a a* stomates.—E. A highly magnified segment of a transverse section of a tendril; *d d* cuticle; *a a* a row of elastic cords of vessels protected by woody fibre next the outside; *b*, *f*, and *c*, *f*, other elastic cords nearer the centre, *b* being the mouths of vessels, and *f* curved masses of woody tissue.

LETTER XLVII.

THE WATER PLANTAIN TRIBE—THE WATER LILY TRIBE.

Plate XLIX.

I HAVE now almost finished all the details about which I propose to occupy you, and so far as systematic Botany is concerned, I see no great object in pursuing the subject further. Indeed, to extend our correspondence much, would change the aim with which it was commenced; and you would be studying a long dissertation upon the Natural System of classification, instead of an introductory account of its elements.

There are, however, two natural orders of aquatic plants, both of which are common in this country, and about which a few remarks may be made with some advantage to you. The first is the WATER PLANTAIN TRIBE, the other the WATER LILY TRIBE.

The *Water Plantain* (Alisma Plantago), and the *Arrow-head* (Sagittaria sagittifolia), are two herbaceous plants, inhabiting the sides of ditches and ponds all over England. In most respects they are alike in the structure of their parts of fructification, differing principally in the latter having more stamens than the former, and these organs in different flowers from

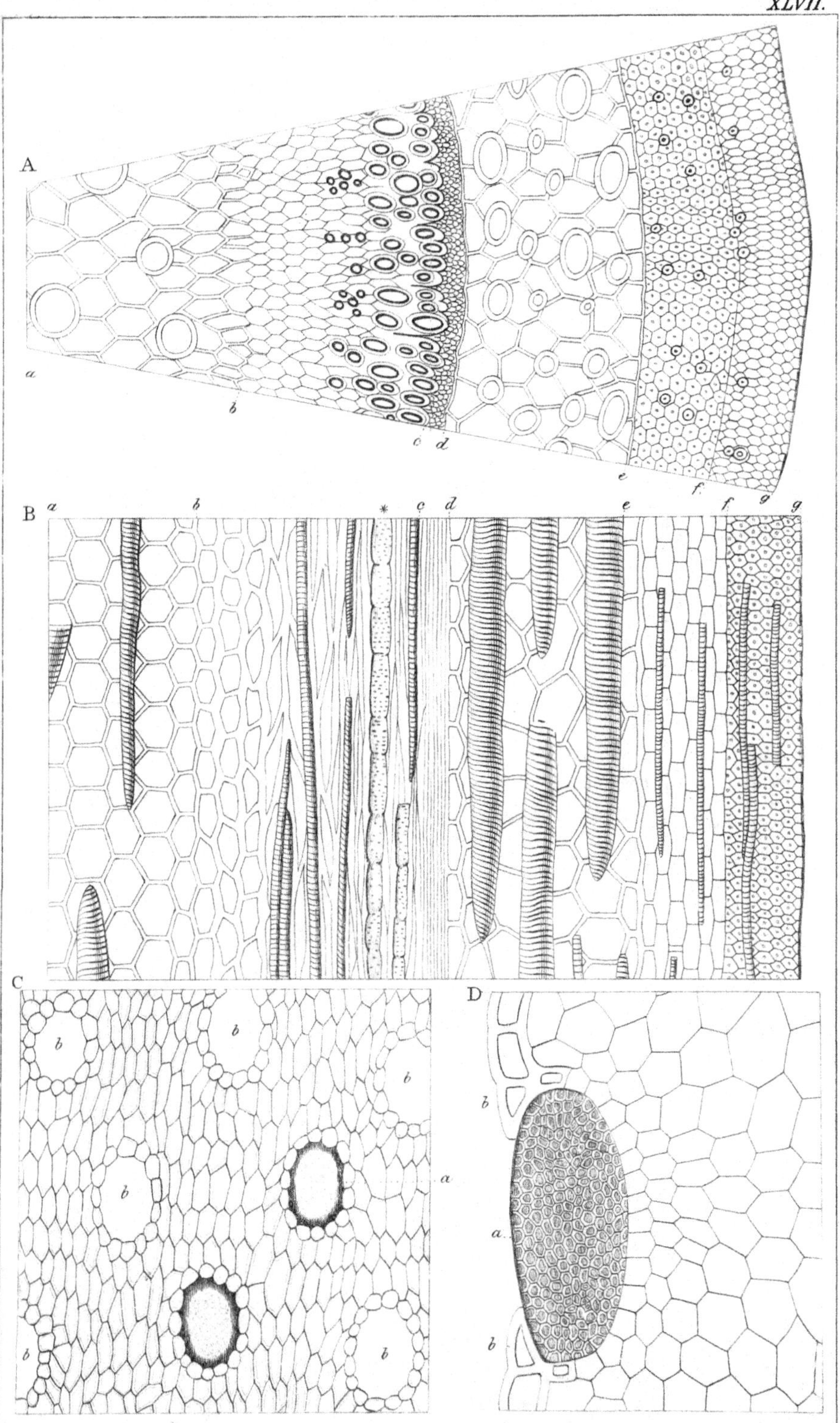

Anatomy of Nepenthes

the pistils. As the Water Plantain is the commoner of the two, let us look at it.

It has oblong, heart-shaped, pointed leaves, marked with about seven ribs, connected by transverse, oblique, forking and branching veins. The flowers are arranged in a loose, whorled, branching panicle (*Plate* XLIX. *fig.* 1.), at the base of each of whose whorls stand a few brown or green ovate bracts. The flowers (*fig.* 2.) have a calyx of three, green, permanent, blunt, parallel-veined sepals, and three delicate pink, or white, roundish, toothed petals.

There are six stamens placed in a very unusual manner, two opposite each sepal (*fig.* 3.); so that in this part of the fructification, Alisma is in a state that cannot be reconciled with the laws of structure before laid down. Upon a more minute examination, however, you will find a small round gland (*fig.* 3. *a.*) at the base of each sepal, and between each pair of stamens; this is obviously a rudimentary stamen, the number of which is thus increased to nine. But still the three stamens that ought to be placed opposite the petals are absent; and they must be considered altogether wanting; the six perfect stamens will belong to two succeeding whorls; so that, in reality, the flower of Alisma, although containing six stamens, or two whorls only, must be considered to be constructed upon a plan of twelve stamens in four whorls, of which the outer is rudimentary, the second deficient, and the two others consolidated into a single whorl; or the scheme of suppression of parts will be expressed thus:—

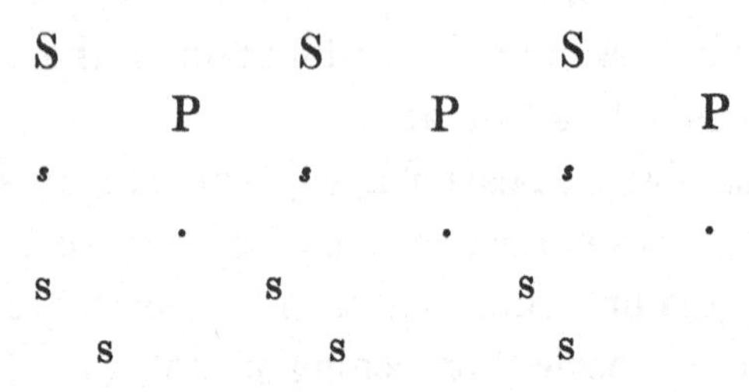

This is of more importance for you to know than you would at first suspect; for it indicates that Alisma, although formed with only six stamens, has a tendency to produce twelve, and hence that it may belong to a tribe, the prevailing number of whose stamens is twelve, or even more; and such is really the fact. Even in Alisma itself, the stamens are in other species nine, twelve, or even more; and in Arrow-head they are in all cases very numerous. Had the six stamens of Alisma belonged to the two first whorls, you would have had no reason to suppose, that although hexandrous, it might have immediate polyandrous affinities.

The ovaries of the Water Plantain are about twenty-four (that is, eight times three), arranged in a somewhat triangular manner; they are quite distinct from each other (*fig.* 4.), and consist of a single cell, from one side of the top of which the style arises in the form of a curved horn, the upper end of which is broken up into a stigma (*fig.* 4. *a.*). There is one ovule (*fig.* 5. *a.*) attached to the bottom of the cell, by a curved stalk.

The fruit (*fig.* 6.) is a triangular head of dry, one-seeded nuts, furrowed at the back, and marked with the base of the style on one side (*fig.* 7. *a.*).

From what has now been stated, can you tell whether this plant is an Exogen or an Endogen? Its leaves are in some measure those of both classes; and not exactly of either. The parallel ribs and netted intervals are, of the two, most like those of an Exogen. The branched *verticillate* inflorescence is most common in Exogens; but then it occurs continually among grasses. The ternary flowers are those of Endogens, but, again, there are many cases among Exogens where the ternary structure also exists: as in the Hepatica which is a Ranunculaceous plant. So far, therefore, as the structure of those parts you have been able to examine is concerned, the evidence seems pretty well balanced.

Perhaps affinity may settle the point. What is Alisma most like? You have no where seen in Endogens an example of numerous carpels and stamens; six, or three, or fewer, having been the prevailing number. We do not, therefore, seem likely to find a parallel in that class. Turn to Exogens, and especially to those which have numerous hypogynous stamens and carpels; and the memory immediately rests upon the Crowfoot Tribe. In that natural order, although the leaves are usually veined in the most legitimately Exogenous manner, yet in some, in the water species in particular, such as the common *tongue-leaved Crowfoot* (Ranunculus Lingua), the veins are disposed upon a plan strikingly similar to that of Alisma; in *Pilewort,* which is a species of Crowfoot (Ranunculus Ficaria), there are only three sepals; and in the *Mousetail* (Myosurus minimus), the stamens fluctuate between five and twenty. The fruit of the Crowfoot Tribe often consists, as you know, of a considerable number of little, one-seeded, closed nuts, with an oblique style at the point; in short, in all these, and some other respects, Alisma is so like a Crowfoot, that it might actually be referred to the tribe of that name by any but a very cautious observer. The principal objection to its being placed in the Crowfoot Tribe, lies here; it is only now and then in Ranunculaceous plants that the number three occurs, and where it does exist, it is confined to the sepals or the petals, and is not found in the stamens or carpels: but in Alisma it occurs throughout every part; in the former, therefore, it may be regarded as an occasional deviation from a rule, while in the latter, it must be looked upon as the rule itself. In fact, the seed of Alisma, which in all these cases is the court of final appeal, shews that Alisma, is in reality, an Endogen.

If you open one of the nuts, you will find the seed standing erect (*fig.* 8.), and containing a monocotyledonous embryo, curved upon itself into the form of a horse-shoe.

The result of this examination shews how necessary it is, in doubtful points, to weigh and balance every thing that can be observed, and not to decide without the most careful investigation. In this case there was no real difficulty in arriving at the truth; it was only care and attention that were required.

The *white Water-Lily* (Nymphæa alba), although an aquatic like the Alisma, is in some respects very different. I select it as another case where a little attention to the rule of evidence in Systematic Botany is required, in order to form a correct judgment. The stem of this plant affords no precise character, either one way or other, as between Exogens and Endogens. Its leaves, moreover, are referable, as much to the type of the one as of the other. Its flowers (*Plate* XLIX. 2. *fig.* 1.) consist of about twenty-five, thickish, oblong leaves, of a dazzling white colour, and the five external ones are more or less green at the back, in representation of a calyx; these leaves grow gradually smaller and smaller towards the centre, till at last their points become callous and yellow; at length bear a pair of short, anther-lobes, in the room of the yellow callosity (*fig.* 3.); these again narrow into straps, having more perfect anthers at the points (*fig.* 4.), and finally, next the ovary, shorten, diminish, and produce less perfect anthers. What I have called anther-bearing petals, are obviously stamens. Do not suppose that in this respect the Water-Lily offers an exception to general rules; in all cases the stamens are nothing but contracted and altered petals provided with anthers; only in the Water-Lily the transition is gradual and apparent, in others, it is too abrupt to be perceived. The number of the stamens is about fifty, but it is not fixed, nor indeed easily ascertained.

The ovary is in a curious state (*fig.* 2.); instead of being either altogether free, or altogether united with the calyx, it has the lower floral leaves free from it, and the upper united with it, so that the anther-bearing petals or stamens grow from just below the stigmas. It has ten or eleven cells, the partitions of which are covered all over with ovules (*fig.* 5.), and the same number of orange-yellow stigmas, which spread away from the centre, like the rays of a poppy-head, to which they bear no little resemblance.

Is this plant an Exogen or an Endogen? Its leaves and stems afford no satisfactory information, and its habit, numerous stamens and carpels, would lead one to think that it bears the same relation to Alisma, as the Poppy to a Crowfoot. But the manifest tendency to the number *five* in the flowers of this plant, is fatal to the supposition; had the tendency been to *four*, the evidence would have still been inconclusive, for four does sometimes occur in the flowers of Endogens; but five, *never*. Therefore, without searching for the seed, the Water-Lily might be confidently considered a polypetalous Exogen; a conclusion confirmed by the seed, which is a little dicotyledonous body, lying in a bag, on the outside of a quantity of farinaceous albumen.

Besides this species, the *yellow Water-Lily* (Nuphar lutea) is extremely common in ponds. Take care, however, that you do not mistake for it the *Floating Buck-bean* (Villarsia nymphæoides), which is a monopetalous plant, belonging to an out-lying portion of the Gentian Tribe.

EXPLANATION OF PLATE XLIX.

I. The Water-Plantain Tribe.—1. A portion of the whorled panicle of *common Water-Plantain* (Alisma Plantago).—2. A complete flower.—3. A calyx, stamens and pistil; *a a* the sepaline glands, or rudimentary stamens.—4. A carpel; *a* its recurved stigmatic face.—5. A section of the ovary, shewing the ovule *a*, elevated on its curved stalk.—6. A fruit.—7. One of the nuts much magnified; *a* the remains of the style.—8. A vertical section of the nut, shewing the seed with its horse-shoe embryo: *a* the base of the style.

II. The Water-Lily Tribe.—1. A flower of the *white Water-Lily* (Nymphæa alba).—2. A vertical section of the pistil, from which the petals, &c. have been cut away; *a a* first transition from petals to stamens; *b* perfect stamens; *c* diminished stamens.—3. A view of the front of a transition petal.—4. A complete stamen.—5. A transverse section of the ovary, with the ovules adhering to all the faces of the dissepiments.

LETTER XLVIII.

THE RIPE FRUIT OF A MANGO.

Plate L.

I SEND you a beautiful drawing, by Mr. Francis Bauer, of the fruit of a Mango, a delicious tropical fruit, which has occasionally been brought to perfection in the hothouses of England, but which is better known in Europe in the form of a pickle. My object in placing the drawing in your hands is to shew you, by its means, something more of the internal structure of a fruit and a seed than you yet possess.

You must remember, that the type of all fruit is the carpel; that all carpels are formed upon one common plan, modified indeed to a great extent, by excessive growth, solidification, attenuation, or the like; and that fruits of every description are composed of one or more carpels, distinct or consolidated, and more or less altered by causes of the same nature as those which affect each separate carpel. So that, to understand the connection that exists between the parts of one ripe carpel, is to possess a standard, to which the peculiarities of all other carpels may be reduced. Nothing more instructive than the Mango can be taken.

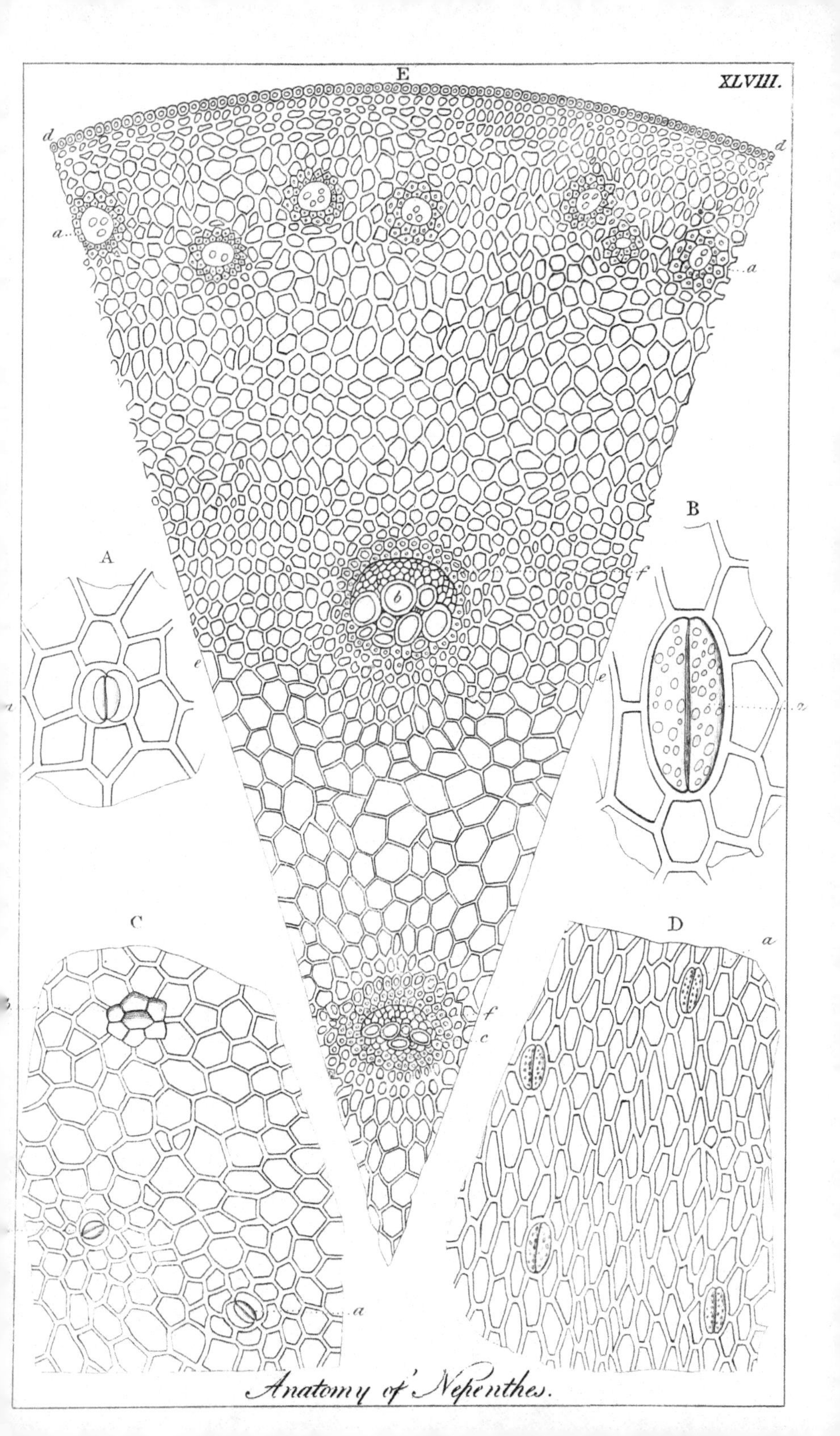

Anatomy of Nepenthes.

The Mango (*fig.* 1.) is an oblong, rather kidney-shaped fruit, composed of an external succulent flesh (*fig.* 2. and 2. * *b.*), adhering to a fibrous woody shell (*fig.* 2. and 2. * *a.*), lined by a hard, homogeneous, brittle crust (*fig.* 2. * *e.*); the whole enclosing the seed (*fig.* 2. * *f.*).

The flesh, shell, and crust, taken together, are the pericarp. They are connected by a prodigious multitude of fibres, which pass from the shell into the flesh, where they lose themselves. The flesh and crust are a continuation of the bark of a branch (*fig.* 2. *d.*), the shell of the wood (*fig.* 2. *c.*), in the organic qualities of which they respectively participate.

The pericarp is theoretically analogous to a leaf rolled inwards, till its edges touch and grow together, so as to form a hollow case. The flesh is analogous to the parenchyma of the lower surface of the leaf, the shell to the veins, the crust to the parenchyma of the upper surface. The parenchyma of the leaf is an extension of the bark of the branch, and the veins of the wood, in the organic qualities of which they likewise participate.

In a leaf the veins convey liquid food from the wood, and deposit it in the parenchyma, where it is digested and altered, and whence it is slowly filtered back into the bark of the branch, which it descends. In the Mango fruit the liquid food is conveyed from the wood into the pericarp by the fibres of the shell, which pour it forth by their thousand mouths into the parenchyma, to be therein digested and altered; but in consequence of the narrowness of the stalk, the cells through which it would have to filter are soon choked up, and then the altered food is forced to accumulate in the parenchyma (*fig.* 2. and 2.* *b.*). Being thus arrested in its course, it swells the tissue in which it lies, becomes more and more changed by constant exposure to light and air, till at last the succulent flesh of the Mango is the result.

As to the parenchyma of the inside of the pericarp, as it is cut off by the shell from all communication with the flesh, and is continually pressed upon by the seed as it grows, being thus jammed as it were between the shell and the seed, it is not unnatural that it should become so hard and solid as we find it.

The seed is attached to the bottom of the pericarp by a broad space (extending from *h* to *e* in *fig.* 3.), and stands erect in the cavity. It has two distinct skins, one of which (*fig.* 3. *b.*) is thin, pale, membranous and loose, the other and inner (*fig.* 3. *c.*) thicker, darker coloured, and fitting close to the embryo.

The inner skin does not grow from the same part of the pericarp as the outer, but springs from the top of a cord which arises obliquely from one side of the base (*fig.* 3. *d.* and 4. *c.*). From its junction with the inner coat to a small depression upon the edge (*fig.* 3. *g.*), the cord throws out veins which, taking a curved direction, and following the form of the embryo, fill the whole of the inner coat with a network of vessels.

The cord alluded to is the raphe, the depression upon the edge of the seed the centre of the chalaza, and in the eyes of physiologists the true organic apex of the seed. It is obvious, therefore, that in this case the organic apex, and the apparent apex, are far from corresponding; and this is a very common occurrence.

The use of the vessels of the chalaza is doubtless to convey from the junction of the pericarp and branch (*fig.* 3. *f.*) the nutritious fluids required to enable the embryo to develope, and to change, from an opaque speck floating in jelly, to a large almond-like kernel.

The embryo is a large almond-like kernel (*fig.* 4. *b.*), composed of two plano-convex cotyledons, curved almost into the form of a kidney, and adhering by a point indicated externally by the small conical radicle (*fig.* 4. *a.*).

If you cut off the cotyledons, so as to get the radicle and plumule small enough to be conveniently magnified about four times, you will see that those two parts form a centre or axis of growth represented by two cones, of which the radicle (*fig.* 5. *a.*), lying in a niche of the cotyledons on the outside, is one, and the plumule (*fig.* 5. *b.*), enclosed between the bases of the cotyledons, is another. The cotyledons grow to the axis by a narrow space (*fig.* 5. *c.*).

The plumule (*figs.* 6. and 7.) is terminated at its point by four extremely minute leaves, crossing or alternating with each other in opposite pairs. Of these plumular leaves, the larger pair (*a. a.*) is external, and partly overlies the smaller (*b.*). The cotyledons themselves, which are larger still, cross or alternate with the outer pair of plumular leaves.

Such would be the position of any three pairs of opposite leaves upon a branch, as you may see by a Laurustinus, or a Sycamore tree; and hence they are all, cotyledons and plumular scales, considered rudimentary, or incompletely formed leaves.

If you can only understand that all fruits whatsoever are either multiplications of that of the Mango, with the addition perhaps of several seeds, and such alterations as I have already spoken of (p. 214), you may form a correct physiological notion of the essential parts of all theories concerning fruits and seeds. For the details relating to so exceedingly curious a subject, I must refer you to any very recent Introductions to Botany, in which the science is treated philosophically.

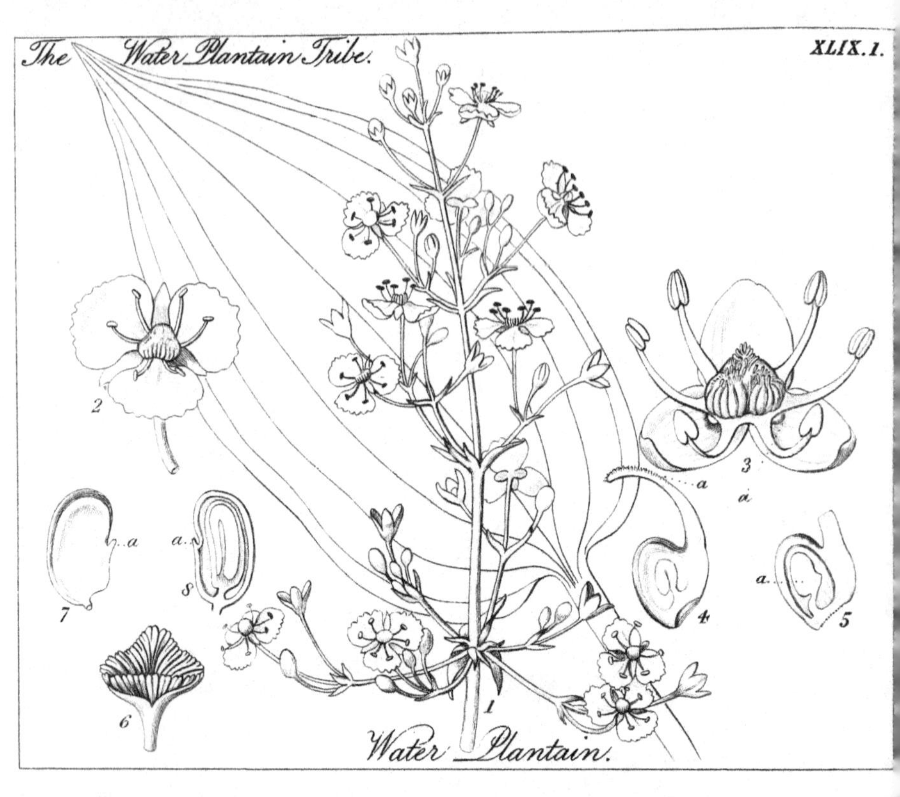
The Water Plantain Tribe.
XLIX. 1.
Water Plantain.

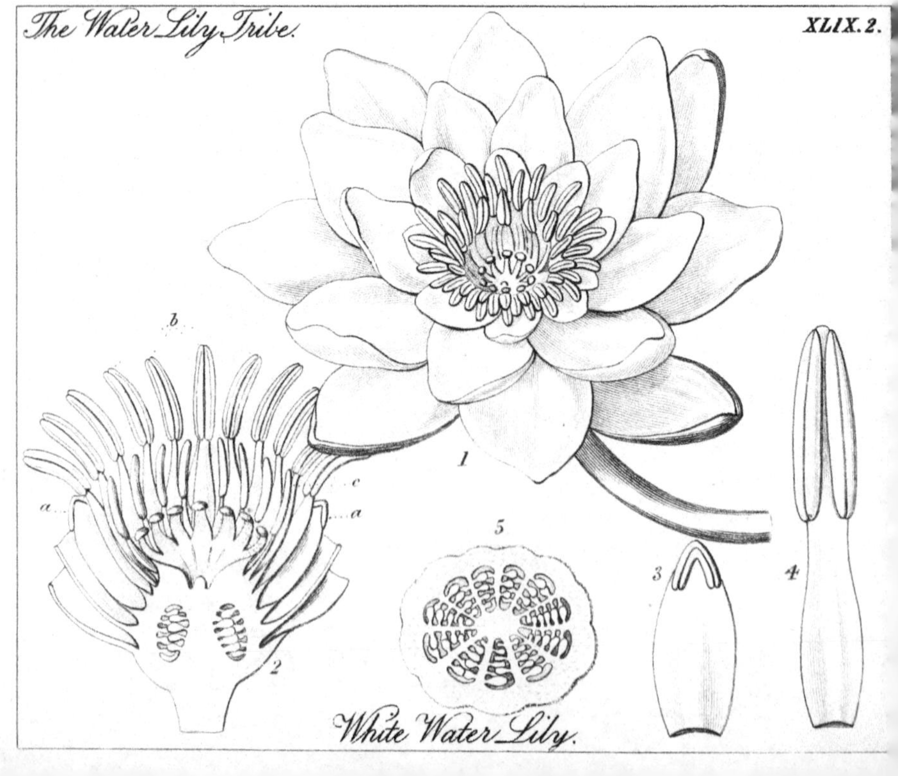
The Water Lily Tribe.
XLIX. 2.
White Water Lily.

LETTER XLIX.

A SYSTEMATIC ARRANGEMENT OF PLANTS, ACCORDING TO THEIR NATURAL RELATIONS, OR SUMS OF RESEMBLANCE.

All that you have learned of the vegetable kingdom has been designedly desultory and unmethodical. My object has been not to engage your attention by explaining to you any particular system, but rather to store your mind with the facts upon which all systems must rest.

But as all systems of arrangement must be unintelligible to those unacquainted with details, so on the other hand must the most copious and well considered details be deprived of a great part of their value, if they are not so arranged as to illustrate and explain each other, as well as to be found whenever the memory seeks for them.

I shall, therefore, without further preface, give you in this letter a sketch of an arrangement of the commoner Natural Orders of plants, according to their resemblances; leaving you to make out the final distinctions between them by such means as you now possess; premising only, that throughout the whole of this compendium I have used the word *tribe*, as an equivalent for what is more generally termed a *natural order*.

There are five CLASSES into which all plants may be divided; namely—

I. EXOGENS, or DICOTYLEDONS; netted-leaved flowering plants, with two or more cotyledons to their embryo, and seeds enclosed in a seed-vessel.

II. GYMNOSPERMS; parallel-veined or fork-veined flowering plants, with two or more cotyledons to their leaves, and seeds formed without the protection of a seed-vessel.

III. ENDOGENS, or MONOCOTYLEDONS; parallel-veined flowering plants, with only one cotyledon.

IV. RHIZANTHS; leafless parasitical flowering plants, with no cotyledons.

V. ACROGENS, or ACOTYLEDONS; plants having no true flowers that can be distinguished, and no cotyledons.

Each class is subdivided according to special rules, and must be treated of separately.

Class I. EXOGENS.

The SUBCLASSES are three; namely—

1. POLYPETALOUS plants; in which the petals are all distinct.
2. MONOPETALOUS plants; in which the petals are united into a tube.
3. *Incomplete* plants; in which there are no petals, and very often not even a calyx.

Each of these subclasses may be again subdivided into *groups*, as follows:—

Subclass I. Polypetalous Plants.

The *groups* are seven; namely—

1. *Albuminous;* a very *minute embryo* in the midst of a large quantity of albumen. This group is separated from the following, because of its remarkable character, and may comprehend all the modifications of structure by which the six following groups are known. In none of the latter is there an embryo much smaller than the albumen; so that, in analyzing the subclass, the student may divide it into two parts, one consisting of the albuminous group exclusively, and the other of the six other groups.
2. *Epigynous;* a large embryo; an *inferior ovary;* the placentation not parietal; the carpels consolidated; the calyx in a perfect whorl.
3. *Parietose;* a large embryo; an inferior or superior ovary indifferently; the *placentæ parietal;* the carpels ccnsolidated; the calyx in a perfect whorl.
4. *Calycose;* a large embryo; a superior ovary; the placentæ not parietal; the carpels consolidated or not; the *calyx in a broken whorl.*
5. *Syncarpous;* a large embryo; a superior ovary; the placentæ not parietal; the *carpels consolidated; the calyx in a perfect whorl.*
6. *Gynobaseous;* a large embryo; a superior ovary; the *cells* of which are *placed obliquely round a conical centre,* and do not exceed five in number;

carpels consolidated or distinct; calyx in a perfect whorl.

7. *Apocarpous;* a large embryo; a superior ovary; *carpels distinct,* and not oblique, if five in number; calyx in a perfect whorl.

Each group is further subdivided into smaller clusters, called *Alliances;* but as you are not acquainted with a sufficient quantity of plants to appreciate such refinements, I shall in this and the succeeding classes simply place the orders you have studied, and a very few others, in little clusters under each of the foregoing groups: adding to their English names their more exact scientific denominations.

Group 1. *Albuminous.*

a. The Crowfoot Tribe (Ranunculaceæ), Plate I. 1.
The Poppy Tribe (Papaveraceæ), Plate I. 2.
The Fumitory Tribe (Fumariaceæ).
The Water Lily Tribe (Nymphæaceæ), Plate XLIX. 2.

b. The Nutmeg Tribe (Myristicaceæ).
The Magnolia Tribe (Magnoliaceæ), Plate XXVI. 1.
The Anona Tribe (Anonaceæ).
The Dillenia Tribe (Dilleniaceæ).

c. The Umbelliferous Tribe (Apiaceæ or Umbelliferæ), Plate II. 1.
The Aralia Tribe (Araliaceæ).

d. The Gooseberry Tribe (Grossulaceæ), Plate XXVII. 1.
The Escallonia Tribe (Escalloniaceæ).

e. The Barberry Tribe (Berberaceæ), Plate XXVI. 2.
The Vine Tribe (Vitaceæ), Plate XXVII. 2.
The Pittosporum Tribe (Pittosporaceæ), Plate XXVIII. 1.
The Francoa Tribe (Francoaceæ).
The Sundew Tribe (Droseraceæ), Plate XXXIII. 2.
The Sidesaddle-flower Tribe (Sarraceniaceæ).

Group 2. *Epigynous.*

a. The Evening Primrose Tribe (Onagraceæ), Plate III. 1.
The Enchanter's Nightshade Tribe (Circæeæ).
The Combretum Tribe (Combretaceæ).
The Melastoma Tribe (Melastomaceæ).
The Myrtle Tribe (Myrtaceæ), Plate III. 2.
The Syringa Tribe (Philadelphaceæ).

b. The Dogwood Tribe (Cornaceæ).
The Miseltoe Tribe (Loranthaceæ).

c. The Gourd Tribe (Cucurbitaceæ), Plate XXX. 2.
The Loasa Tribe (Loasaceæ).
The Cactus Tribe (Cactaceæ), Plate XXX. 1.
The Fig-Marigold Tribe (Mesembryaceæ), Plate XXXI. 2.
The Begonia Tribe (Begoniaceæ), Plate XXXI. 1.

Group 3. *Parietose.*

a. The Cruciferous Tribe (Brassicaceæ or Cruciferæ), Plate IV. 1.
The Caper Tribe (Capparidaceæ), Plate XXIX. 2.
The Mignonette Tribe (Resedaceæ), Plate XXIX. 1.

b. The Violet Tribe (Violaceæ), Plate IV. 2.
The Frankenia Tribe (Frankeniaceæ).

c. The Passion-flower Tribe (Passifloraceæ), Plate V. 1.
The Turnera Tribe (Turneraceæ).

Group 4. *Calycose.*

a. The Guttiferous Tribe (Clusiaceæ or Guttiferæ).
The Tutsan Tribe (Hypericaceæ), Plate V. 2.

b. The Tea Tribe (Ternströmiaceæ).

c. The Maple Tribe (Aceraceæ).
The Horse-chesnut Tribe (Æsculaceæ), Plate XXXVI. 1.
The Soap-berry Tribe (Sapindaceæ).
The Milk-wort Tribe (Polygalaceæ), Plate XXVIII. 2.

d. The Rock Rose Tribe (Cistaceæ), Plate XXXII. 2.
The Flax Tribe (Linaceæ), Plate XXXIX. 1.

Group 5. *Syncarpous.*

a. The Lythrum Tribe (Lythraceæ), Plate XXXII. 1.
The Mallow Tribe (Malvaceæ), Plate VI. 1.

The Sterculia Tribe (Sterculiaceæ).
The Linden Tribe (Tiliaceæ).

b. The Orange Tribe (Aurantiaceæ), Plate VI. 2.

c. The Buckthorn Tribe (Rhamnaceæ), Plate XXXVIII. 1.
The Euphorbia Tribe (Euphorbiaceæ), Plate XXXVIII. 2.
The Crowberry Tribe (Empetraceæ).
The Celastrus Tribe (Celastraceæ).
The Bladder-nut Tribe (Staphyleaceæ).
The Malpighia Tribe (Malpighiaceæ).

d. The Lychnis Tribe (Silenaceæ), Plate VII. 1.
The Chickweed Tribe (Alsinaceæ).
The Purslane Tribe (Portulacaceæ), Plate VII. 2.
The Tamarisk Tribe (Tamaricaceæ), Plate XXXIII. 1.
The Knot-Grass Tribe (Illecebraceæ).

Group 6. *Gynobaseous.*

a. The Rue Tribe (Rutaceæ), Plate XXXIX. 2.
The Bean-Caper Tribe (Zygophyllaceæ).
The Yellow-wood Tribe (Xanthoxylaceæ)

b. The Geranium Tribe (Geraniaceæ), Plate II. 2.
The Balsam Tribe (Balsaminaceæ).
The Nasturtium Tribe (Tropæoleæ).
The Wood-sorrel Tribe (Oxalidaceæ).

c. The Coriaria Tribe (Coriariaceæ).

d. The Limnanthes Tribe (Limnanthaceæ).

Group 7. *Apocarpous.*

a. The Rose Tribe (Rosaceæ), Plate VIII. 1.
The Apple Tribe (Pomeæ).
The Almond Tribe (Amygdaleæ).
The Burnet Tribe (Sanguisorbeæ).
The Pea Tribe (Leguminosæ), Plate VIII. 2.
The Carolina Allspice Tribe (Calycanthaceæ).

b. The Saxifrage Tribe (Saxifragaceæ), Plate XXXVII. 2.
The Bauera Tribe (Baueraceæ).
The Houseleek Tribe (Crassulaceæ), Plate XXXVII. 1.

Subclass II. MONOPETALOUS PLANTS.

The *groups* are five; namely—

1. *Polycarpous;* ovary of *several carpels,* either distinct or consolidated, and *never inferior,* except in one case; fruit never bony and nut-like.
2. *Epigynous;* ovary of *several carpels,* either distinct or consolidated, and *inferior in all cases.*
3. *Aggregose;* ovary of *one carpel only,* and that either superior or inferior.
4. *Nucamentous;* ovary of two or more carpels, which change to *bony nuts* or seed-like pericarps, and are never inferior.
5. *Dicarpous;* ovary of *two carpels,* which are always *superior,* and do not change to bony nuts or seed-like pericarps.

The commoner natural orders belonging to these groups, are as follows:—

Group 1. *Polycarpous.*

a. The Winter-green Tribe (Pyrolaceæ).
The Monotropa Tribe (Monotropaceæ).
The Heath Tribe (Ericaceæ), Plate XII. 1.
The Bilberry Tribe (Vaccinaceæ).
The Epacris Tribe (Epacridaceæ), Plate XLII. 2.

b. The Primrose Tribe (Primulaceæ), Plate XLII. 1.
The Myrsien Tribe (Myrsinaceæ).
The Holly Tribe (Aquifoliaceæ).
The Styrax Tribe (Styraceæ).

c. The Nolana Tribe (Nolanaceæ).
The Bindweed Tribe (Convolvulaceæ), Plate XII. 2.
The Dodder Tribe (Cuscutaceæ).
The Greek Valerian Tribe (Polemoniaceæ), Plate XLIII. 1.
The Diapensia Tribe (Diapensiaceæ).
The Hydrolea Tribe (Hydroleaceæ).

Group 2. *Epigynous.*

a. The Lobelia Tribe (Lobeliaceæ).
The Harebell Tribe (Campanulaceæ), Plate XIV. 1.
The Stylidium Tribe (Stylidiaceæ).
The Goodenia Tribe (Goodeniaceæ).

b. The Coffee Tribe (Cinchonaceæ).
The Honey-suckle Tribe (Caprifoliaceæ), Plate XIV. 2.
The Madder Tribe (Galiaceæ, or Stellatæ), Plate XLIV. 1.

Group 3. *Aggregosæ.*

a. The Composite-flowered Tribe (Asteraceæ or Compositæ), Plate XVII. 1.
The Scabious Tribe (Dipsaceæ), Plate XLIV. 2.
The Valerian Tribe (Valerianaceæ).
The Brunonia Tribe (Brunoniaceæ).

b. The Rib-grass Tribe (Plantaginaceæ), Plate XVII. 2.
The Globularia Tribe (Globulariaceæ).
The Leadwort Tribe (Plumbaginaceæ).

Group 4. *Nucamentous.*

a. The Water-leaf Tribe (Hydrophyllaceæ).
The Borage Tribe (Boraginaceæ), Plate XV. 1.

b. The Mint Tribe (Lamiaceæ, or Labiatæ), Plate XVI. 1.
The Vervain Tribe (Verbenaceæ).
The Myoporum Tribe (Myoporaceæ).

Group 5. *Dicarpous.*

a. The Trumpet-flower Tribe (Bignoniaceæ), Plate XLIII. 2.

b. The Justicia Tribe (Acanthaceæ).
The Butterwort Tribe (Lentibulaceæ).
The Gesnera Tribe (Gesneraceæ).
The Broom-Rape Tribe (Orobanchaceæ).
The Foxglove Tribe (Scrophulariaceæ), Plate XVI. 2.

c. The Nightshade Tribe (Solanaceæ), Plate XV. 2.

d. The Gentian Tribe (Gentianaceæ), Plate XIII. 1.
The Wormseed Tribe (Spigeliaceæ).
The Apocynum Tribe (Apocynaceæ),
The Asclepias Tribe (Asclepiadaceæ), Plate XLV. 2.

e. The Olive Tribe (Oleaceæ), Plate XIII. 2.
The Jasmine Tribe (Jasminaceæ), Plate XLV. 1.

Subclass III. INCOMPLETE PLANTS.

The *groups* are five; namely—

1. *Rectembryous;* calyx imperfect; *embryo straight.*
2. *Achlamydeous;* calyx entirely *wanting.*
3. *Tubiferous;* calyx *tubular,* often resembling a corolla; embryo straight; ovary usually one-celled.
4. *Columnous;* calyx perfect; *ovary 3-6-celled;* embryo straight.
5. *Curvembryous;* calyx perfect; *embryo curved* like a horse-shoe.

The common natural orders belonging to these groups are as follows:—

Group 1. *Rectembryous.*

a. The Oak Tribe (Corylaceæ or Cupuliferæ), Plate X. 2.
The Birch Tribe (Betulaceæ).
The Garrya Tribe (Garryaceæ).

b. The Nettle Tribe (Urticaceæ), Plate XI. 1.

The Elm Tribe (Ulmaceæ).
The Gale Tribe (Myricaceæ).
The Walnut Tribe (Juglandaceæ), Plate XXXVI. 2.

Group 2. *Achlamydeous.*

a. The Saururus Tribe (Saururaceæ).
The Pepper Tribe (Piperaceæ).
b. The Willow Tribe (Salicaceæ), Plate XI. 2.
The Plane Tribe (Platanaceæ).
c. The Callitriche Tribe (Callitrichaceæ).

Group 3. *Tubiferous.*

a. The Oleaster Tribe (Elæagnaceæ).
The Mezereum Tribe (Thymelaceæ), Plate XLI. 1.
The Protea Tribe (Proteaceæ), Plate IX. 1.
b. The Cinnamon Tribe (Lauraceæ), Plate XLI. 2.

Group 4. *Columnous.*

The Birthwort Tribe (Aristolochiaceæ), Plate XLVI. 1.
The Nepenthes Tribe (Nepenthaceæ).

Group 5. *Curvembryous.*

a. The Amaranth Tribe (Amaranthaceæ), Plate IX. 2.
The Goosefoot Tribe (Chenopodiaceæ), Plate XL. 2.
The Tetragonia Tribe (Tetragoniaceæ).
The Phytolacca Tribe (Phytolaccaceæ).
The Buck-wheat Tribe (Polygonaceæ) Plate XL. 1.

b. The Knawel Tribe (Scleranthaceæ).
The Marvel of Peru Tribe (Nyctaginaceæ), Plate X. 1.

Class II. GYMNOSPERMS.

The Fir Tribe (Pinaceæ or Coniferæ).
The Yew Tribe (Taxaceæ).
The Cycas Tribe (Cycadaceæ).
The Horsetail Tribe (Equisetaceæ).

Class III. ENDOGENS.

There are no subclasses; but there are five principal *groups*, viz.:—

1. *Epigynous; ovary inferior;* stamens and style distinct.
2. *Gynandrous;* ovary inferior; *stamens and style consolidated.*
3. *Hypogynous;* ovary superior; *flowers perfect.*
4. *Spadiceous;* ovary superior; *flowers usually in a spadix,* imperfect, either naked, or composed of a whorl of scales.
5. *Glumose;* ovary superior; *flowers* imperfect, *composed of* imbricated *ribbed bracts.*

Under these groups are arranged the following natural orders:—

Group 1. *Epigynous.*

a. The Ginger Tribe (Zingiberaceæ).
The Arrow-root Tribe (Marantaceæ).
The Banana Tribe (Musaceæ).

b. The Narcissus Tribe (Amaryllidaceæ), Plate XVIII. 1.
The Cornflag Tribe (Iridaceæ), Plate XVIII. 2.
The Pine Apple Tribe (Bromeliaceæ).

Group 2. *Gynandrous.*

The Orchis Tribe (Orchidaceæ), Plate XIX. 2.
The Vanilla Tribe (Vanillaceæ).

Group 3. *Hypogynous.*

a. The Palm Tribe (Palmaceæ).

b. The Lily Tribe (Liliaceæ), Plate XX. 1.
The Asphodel Tribe (Asphodeleæ), Plate XIX. 1.
The Colchicum Tribe (Melanthaceæ).

c. The Spiderwort Tribe (Commelinaceæ).
The Flowering Rush Tribe (Butomaceæ).
The Water Plantain Tribe (Alismaceæ), Plate XLIX. 1.
The Rush Tribe (Juncaceæ), Plate XX. 2.

Group 4. *Spadiceous.*

a. The Arum Tribe (Araceæ), Plate XLVI. 2.
The Acorus Tribe (Acoraceæ).
The Bulrush Tribe (Typhaceæ), Plate XXI. 1.

b. The Naiad Tribe (Naiadaceæ).
The Arrow-grass Tribe (Juncaginaceæ), Plate XXI. 2.
The Duckweed Tribe (Pistiaceæ), Plate XXI. 3.

Group 5. *Glumose.*

The Grass Tribe (Graminaceæ), Plate XXII. 1.

The Sedge Tribe (Cyperaceæ), Plate XXII. 2.

Class IV. RHIZANTHS.

There are no plants of this class either cultivated in gardens, or wild in the North of Europe.

Class V. ACROGENS.

Neither subclasses nor groups are distinguished in this class; the commoner natural orders are—

a. The Fern Tribe (Filicales), Plate XXIII. 1.

The Club-moss Tribe (Lycopodiaceæ), Plate XXIII. 2.

The Moss Tribe (Bryaceæ or Musci), Plate XXIV. 1.

The Jungermannia Tribe (Jungermanniaceæ), Plate XXIV. 2.

The Liverwort Tribe (Marchantiaceæ or Hepaticæ).

b. The Chara Tribe (Characeæ).

c. The Mushroom Tribe (Fungaceæ), Plate XXV. 3.

The Lichen Tribe (Lichenaceæ), Plate XXV. 1.

The Sea-weed Tribe (Algaceæ), Plate XXV. 2.

LETTER L.

AN ARTIFICIAL METHOD OF DISCOVERING WITH CERTAINTY THE NATURAL ORDER TO WHICH A GIVEN PLANT BELONGS.

It is to be supposed that you are by this time well grounded in the distinctions of the commoner Natural Orders of plants; and my last letter will have furnished you with the means of arranging your knowledge in a methodical way. I, therefore, might with this have left you to your own resources in future, or have referred you to the higher systematical works of Botanists, for the means of carrying your inquiries further. But I am so anxious to remove every impediment from your path, that I have prepared for you a set of tables, by means of which you may with certainty discover to what Natural Order any given plant belongs, without being obliged to examine it so minutely as is in some instances necessary in a natural arrangement.

You will, doubtless, have remarked, that some of the distinctions between the groups, as disposed in my last letter, are minute, and difficult to discover; especially those which are taken from the structure of the seed. You will also find, in practice, that there are many exceptions to the characters of the subclasses and groups; for instance, *Virgin's Bower* (Clematis), *Spurge* (Euphorbia), *Mare's-tail* (Hippuris), and *La-*

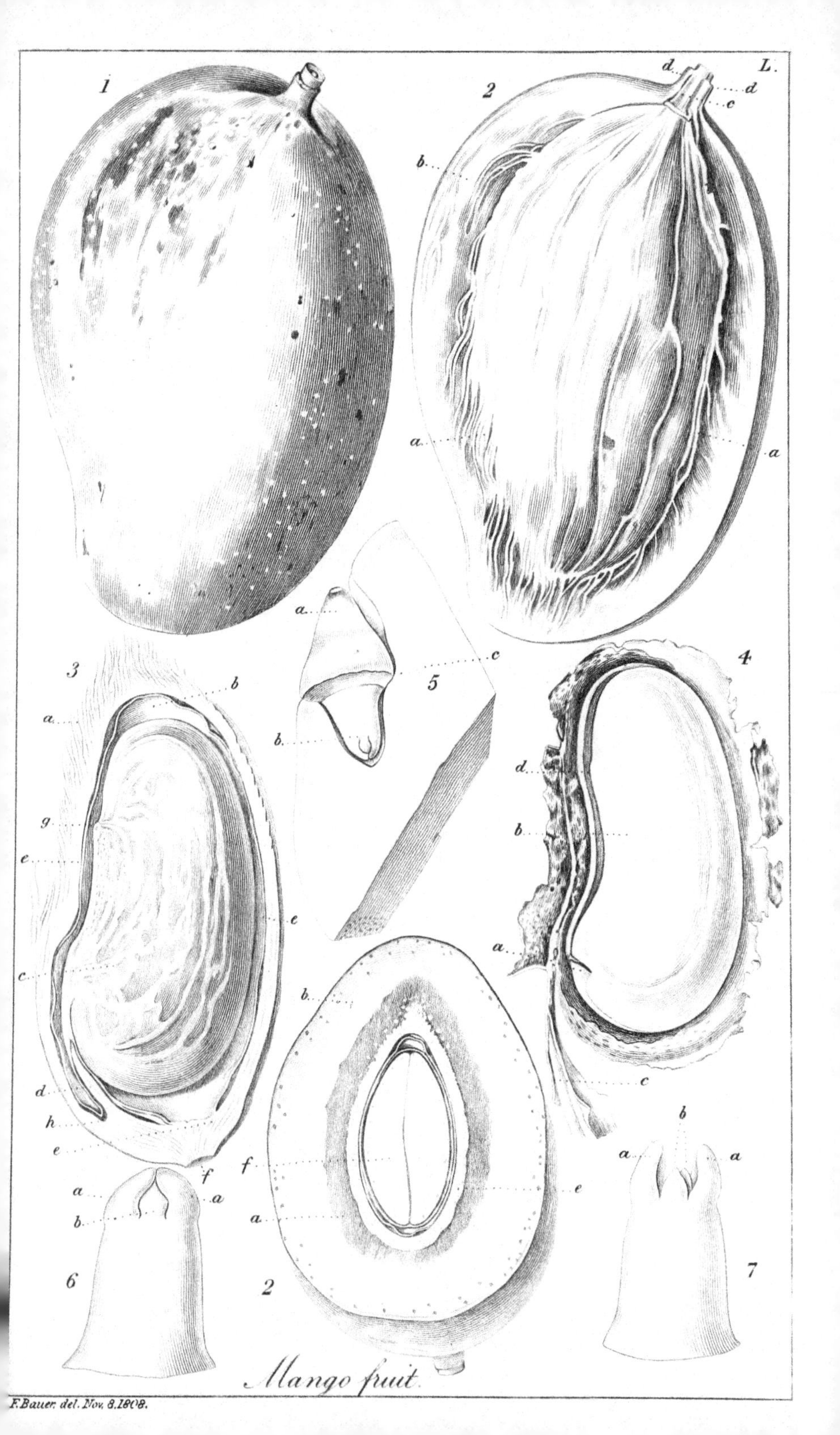
L.
1
2
d
d
c
b
a
a
a
c
5
b
3
b
a
g
e
e
c
d
h
e
f
4
d
b
a
c
b
f
e
a
a
a
b
6
2
b
a
a
7
Mango fruit.
F. Bauer. del. Nov. 8. 1808.

dies'-mantle (Alchemilla), belonging to the Polypetalous subclass of Exogens, have in reality no petals; *Glaux*, belonging to the Monopetalous subclass of Exogens, has no petals; *Correa*, belonging to the Polypetalous subclass of Exogens, has a monopetalous corolla; and so on. No doubt, a Botanist who has had a little experience, overcomes these difficulties easily enough, because he will recognize the plants by the remainder of their structure, and notwithstanding their deviation from the general rule. But, although there exist exceptions to all rules whatsoever, and every person must, therefore, be accustomed to contend with them, whatever the branch of knowledge to which his studies have been directed; yet it must be confessed, that they are always very embarrassing to a beginner, and should be provided against by the best means that can be devised. Therefore, as my parting gift, and an appropriate conclusion to the correspondence that has passed between us, I send you a key, not only to all the Natural Orders of plants you have yet seen, but also to such others as you are at all likely to meet with.

It is only necessary for you to know how to use this key, and I confidently expect you will be at once relieved from all future embarrassment, both in distinguishing the orders themselves, and in guarding yourself against errors arising from exceptional cases. I would, indeed, advise you at first to use your key in all cases whatever, whether of doubt or not; for you will find it give you a habit of examining plants carefully, instead of looking at them superficially.

The principle upon which the key is constructed is

always to contrast characters in pairs, and to refer from one contrast to another, till at last there is nothing left out of which a further contrast can be drawn up ; at that point, where comparison ceases, you ought to find the object of your search. This, which is called the dichotomous analysis, is that, in fact, which the human mind habitually, though unconsciously, employs in all its operations ; and it possesses the great merit of being unerring, provided the comparisons are made with due caution. The best mode of instructing you how to use it, is to select a few examples; first, of plants conformable to the characters assigned to their orders ; and, secondly, of others which offer exceptions to their characters.

The *Pellitory* (Parietaria officinalis), is a plant quite conformable to the characters assigned to its order. Take it as a test. You look to the first pair of characters, or *contrast No.* 1. in the table, and you have no difficulty in deciding that it belongs to " Plants having distinct and visible flowers ;" the No. 2. at the end of that line carries you to *contrast No.* 2. where it agrees with " Leaves not-veined," &c. Then, you proceed to No. 3. where you find that the Pellitory agrees with " Flowers incomplete, that is, having no corolla." You are now referred to No. 97. where you see that your plant corresponds with the character " Calyx present in some kind of state ;" this takes you to No. 105. where you have no difficulty in selecting " Ovary superior," referring to No. 119. as that which suits your plant. At 119, " Leaves with stipules" corresponds with the Pellitory, and thus you reach No. 120. At that point you, of course, take " Flowers unisexual," and so proceed to No. 121. Here you find " Carpels solitary," contrasted with " Carpels more than 1 ;" and it is obvious that your plant belongs to the first; it is, therefore, of the *Nettle Tribe;* and thus you have reached the desired information.

Let Sage (Salvia officinalis) be the next plant for trying the key. Without going over again the words of the earlier parts of the contrast, it is sufficient to say, that No. 1. refers you to No. 2., No. 2. to No. 3., No. 3. to No. 4., No. 4. to No. 157. ; hence we will proceed more carefully. At this point " Ovary superior" sends you to No. 158. ; thence " Flowers irregular" take you to No. 185. ; when there you have " Ovary four-lobed," contrasted with " Ovary undivided," and as your Sage corresponds with the former, it belongs to the *Mint Tribe.*

Now, for a case or two of plants that do not correspond with all the characters assigned to the orders to which they belong. It is here that the tables should be of the greatest use; for the last thing which a student learns, is how to deal with exceptions. *Glaux,* a little coast plant, common on the sandy beach in many places, is a remarkable puzzle; it is destitute of corolla, and yet it belongs to the Monopetalous Primrose Tribe, with which it corresponds in every thing, except the presence of a corolla. You will readily detect its place in the system by the table. As before, No. 1. refers you to No. 2. ; No. 2. to No. 3. ; No. 3. to No. 97., where it corresponds with

"Calyx present in some kind of state," 105.; at that number "Ovary superior", carries you forward to No. 119.; at that point, "Leaves without stipules" refer you to No. 133.; there "Flowers bisexual" take you to No. 134.; thence "Sepals more than two," to No. 135. There you may be stopped by not knowing whether Glaux, with a one-celled ovary, and a free central placenta, belongs to 136. or 144.; but a little reflection will remind you, that such a structure is a consequence of the consolidation of several carpels (see Vol. 2. page 214.), and, consequently, you decide for No. 136.; at that contrast, "Placentas in the axis" correspond with your plant, and you move on to No. 137.; thence by "Number of ovules very great," to No. 140.; then by "Carpels consolidated at the point" to No. 141., whence "Stamens perigynous," carry you to No. 142., where, finally, you have the character "Fruit one-celled," which safely disembarks you in the desired haven, the land of the *Primrose Tribe.*

Another instance, and I have done. *Correa*, a common, and very pretty genus of the Rue Tribe, has its petals united into a tube, so as to seem as if monopetalous, although the plant belongs to a Polypetalous order. This, then, is a great puzzle to a beginner, and a fitting subject by which to try the goodness of the tables. You will first proceed from No. 1. to No. 2.; from No. 2. to No. 3.; and from No. 3. to No. 4. Here, if Correa were conformable to the character of its order, you would proceed to No. 5.; but, as it is monopetalous, and, therefore, unconformable, you take another road, and advance at once to No. 157. From that point, " Ovary superior " leads you to No. 158. ; thence " Flowers regular " to No. 159. ; from that you are conducted by " Ovary lobed " to No. 160. ; where the " dotted leaves " fix your plant with the Rue Tribe, to which it really belongs.

Nothing can be more easy than the use of this table ; and now that you possess it, I confidently expect that you will say in your next letter, " Now you *have indeed* shewn me the way out of my perplexities." Remember only that the table is not contrived to meet all cases whatsoever, for a great many Natural Orders are not even mentioned in it. It is only framed to enable you to master such difficulties as you, as a learner, may be expected to meet with, either in fields or gardens.

TABLE.

1. Plants having distinct and visible flowers . 2
   Plants having no visible flowers . 225

2. Leaves net-veined. Wood in concentric layers 3
   Leaves straight-veined, or feather-veined. Wood not in concentric layers . 205

3. Flowers complete ; that is, having both calyx and corolla . . . 4
   Flowers incomplete ; that is, having no corolla 97

4. Corolla polypetalous ; that is, the petals distinct 5
   Corolla monopetalous ; that is, the petals joined into a tube . . . 157

5. Stamens more than twenty . . 6
Stamens fewer than twenty . . 31

6. Ovary inferior; that is, adhering to the calyx more or less . . . 7
Ovary superior; that is, not adhering at all to the calyx 14

7. Leaves with stipules . *The Apple Tr.*
Leaves without stipules . . 8

8. Carpels more or less distinct from each other . . *The Bauera Tr.*
Carpels wholly combined . . 9

9. Placentas spread over the whole surface of the partitions of the fruit *The Water Lily Tr.*
Placentas confined to the centre or sides of the fruit 10

10. Placentas parietal; that is, adhering to the sides of the fruit . . . 11
Placentas central; that is, growing together in the middle of the fruit . . 12

11. Petals few in number, and different from the sepals . . *The Loasa Tr.*
Petals numerous, and undistinguishable from the sepals . . *The Cactus Tr.*

12. Leaves with little transparent dots *The Myrtle Tr.*
Leaves quite opaque . . 13

13. Petals very numerous *The Fig-Marigold Tr.*
Petals very few (4—5) . *The Syringa Tr.*

14. Leaves with stipules . . 15
Leaves without stipules . . 21

15. Carpels more or less distinct . 16
Carpels wholly consolidated . . 17

16. Stamens hypogynous . *The Magnolia Tr.*
Stamens perigynous . *The Rose Tr.*

17. Æstivation of the calyx imbricated . 18
Æstivation of the calyx valvate . 20

18. Flowers unisexual; that is, stamens in one flower, and pistil in another *The Euphorbia Tr.*
Flowers bisexual; that is, with stamens and pistil in the same flower . . 19

19. Sepals two . . *The Purslane Tr.*
Sepals three or five . *The Rock Rose Tr.*

20. Stamens monadelphous; that is, united with each other in a tube *The Mallow Tr.*
Stamens all distinct . *The Linden Tr.*

21. Carpels more or less distinct . 22
Carpels quite consolidated . . 25

22. Stamens perigynous . *The Rose Tr.*
Stamens hypogynous . . 23

23. Calyx in a broken whorl *The Tutsan Tr.*
Calyx in a perfect whorl . . 24

24. Acrid nauseous herbs *The Crowfoot Tr.*
Aromatic shrubs or trees *The Anona Tr.*

25. Fruit one-celled . . . 26
Fruit many-celled . . . 27

26. Ovary stalked. Sap watery *The Caper Tr.*
Ovary sessile. Sap milky *The Poppy Tr.*

27. Placentas spread over the dissep. *The Water Lily Tr.*
Placentas in the axis of the fruit 28

28. Stigma large broad and peltate *The Sidesaddle Tr.*
Stigma small and simple . . 29

29. Ovary one-celled . *The Purslane Tr.*
Ovary many-celled . . . 30

30. Calyx tubular furrowed. Stamens perigynous *The Lythrum Tr.*
Calyx of three or five leaves in a broken whorl. Stamens hypogynous *The Rock Rose Tr.*

31. Ovary more or less inferior . . 32
Ovary entirely superior . . . 44

32. Leaves with stipules . . . 33
Leaves without any stipules . . 34

33. Flowers unisexual . *The Begonia Tr.*
Flowers bisexual . . *The Buckthorn Tr.*

34. Placentas parietal . . . 35
Placentas in the axis . . . 36

35. Flowers unisexual . *The Gourd Tr.*
Flowers bisexual . *The Currant Tr.*

36. Flowers in umbels . *The Umbelliferous Tr.*
Flowers not in umbels . . . 37

37. Carpels solitary . . *The Combretum Tr.*
Carpels more than one . . . 38

38. Carpels divaricating at point *The Saxifrage Tr.*
Carpels quite parallel and united . 39

39. Æstivation of calyx valvate . . 40
Æstivation of calyx imbricated . . 42

40. Fruit many seeded . *The Evening Primrose Tr.*
Fruit very few seeded . . . 41

41. Stamens opposite the petals . *The Buckthorn Tr.*
Stam alternate with the petals *The Dogwood Tr.*

42. Leaves dotted . . *The Myrtle Tr.*
Leaves not dotted . . . 43

43. Stam. doubled down in flower-bud *The Melastoma Tr.*
Stamens erect . *The Escallonia Tr.*

44. Leaves with stipules . . . 45
Leaves without stipules . . . 60

45. Carpels distinct or solitary . . 46
Carpels consolidated . . . 48

46. Anthers with recurved valves . *The Barberry Tr.*
Anthers with longitudinal valves . 47

47. Fruit a pod . . . *The Pea Tr.*
Fruit a capsule, or little drupe . *The Rose Tr.*

48. Placentas parietal . . . 49
Placentas in the axis . . 51

49. Flowers with filamentous crown *The Passion Fl. Tr.*
Flowers crownless . . . 50

50. Leaves circinate; that is, coiled up, when young . . . *The Sun-dew Tr.*
Leaves straight when young . *The Violet Tr.*

51. Styles distinct to the base . . 52
Styles more or less combined . . 54

52. Flowers unisexual . *The Euphorbia Tr.*
Flowers bisexual . . . 53

53. Petals very minute . *The Knotgrass Tr.*
Petals very obvious . *The Saxifrage Tr.*

54. Æstivation of calyx imbricated . . 55
Æstivation of calyx valvate . . 59

55. Leaves regularly opposite . . 56
Leaves alternate, or only occasionally opposite 57

56. Stem articulated; i. e. separating into distinct pieces at the joints *The Bean-caper Tr.*
Stem continuous *The Bladder Nut Tr.*

57. Calyx in a complete whorl . . . 58
Calyx in a broken whorl . *The Soap-tree Tr.*
Calyx of only two sepals . *The Purslane Tr.*

58. Fruit beaked . . *The Geranium Tr.*
Fruit not beaked . *The Wood Sorrel Tr.*

59. Stamens perigynous . *The Buckthorn Tr.*
Stamens hypogynous . . *The Vine Tr.*

60. Carpels more or less distinct, or solitary 61
Carpels consolidated . . . 68

61. Anthers with recurved valves . *The Barberry Tr.*
Anthers with longitudinal valves . . 62

62. Fruit a legume . . *The Pea Tr.*
Fruit not a legume . . . 63

63. Carpels with hypogynous scales . . 64
Carpels without hypogynous scales . 65

64. One hypog. scale to each carpel *The Houseleek Tr.*
Two hypog. scales to each carpel *The Francoa Tr.*

65. Cal. & cor. undistinguishable *The Carolina Allsp. Tr.*
Calyx and corolla quite different . . 66

66. Herbaceous plants . *The Crowfoot Tr.*
Trees or shrubs . . . 67

67. Cal. and cor. divided into threes *The Anona Tr.*
Cal. and cor. divided into fours *The Coriaria Tr.*

68. Fruit one-celled; if two-celled, then the dissepiment a spurious one . . 69
Fruit with several cells . . 73

69. Stamens tetradynamous; that is, four long and two short . *The Cruciferous Tr.*
Stamens not tetradynamous . . 70

70. Hypogynous, disk large . . 71
Hypogynous, disk absent . . 72

71. Ovary stalked . . . *The Caper Tr.*
Ovary sessile . . *The Mignonette Tr.*

72. Calyx 5-leaved . . *The Turnera Tr.*
Calyx 3 or 4-leaved . . *The Poppy Tr.*

73. Placentas covering the dissep. *The Water-Lily Tr.*
Placentas confined to the axis . 74

74. Styles distinct to the base . . 75
Styles consolidated . . . 80

75. Calyx in a broken whorl . . 76
Calyx in a perfect whorl . . 77

76. Stamens in several parcels *The Tutsan Tr.*
Stam. in a perfect whorl (monadelphous) *Flax Tr.*

77. Carpels each subtended by an hypogynous scale . . *The Houseleek Tr.*
Carpels scaleless . . . 78

78. Carpels 2, divaricating at end *The Saxifrage Tr.*
Carpels more than two, often with a free central placenta . . . 79

79. Calyx tubular . . *The Catchfly Tr.*
Calyx 5-, or 4-parted . *The Chickweed Tr.*

80. Æstivation of calyx imbricated . 81
Æstivation of calyx valvate or open . 96

81.	Sepals in a broken whorl . .	82
	Sepals in a complete whorl . .	85
82.	Fruit splitting into valves	*The Horsechesnut Tr.*
	Fruit not splitting . . .	83
83.	Calyx papilionaceous .	*The Milkwort Tr.*
	Calyx uniform . . .	84
84.	Petals without appendages .	*The Maple Tr.*
	Petals with appendages .	*The Soap-tree Tr.*
85.	Flowers unisexual . . .	86
	Flowers bisexual . . .	87
86.	Leaves dotted .	*The Yellowwood Tr.*
	Leaves heath-like and dotless	*The Crowberry Tr.*
87.	Leaves dotted . . .	88
	Leaves not dotted . . .	89
88.	Fruit a dry capsule . .	*The Rue Tr.*
	Fruit a succulent berry .	*The Orange Tr.*
89.	Flowers irregular . .	*The Balsam Tr.*
	Flowers regular . . .	90
90.	Carpels four or more . . .	91
	Carpels fewer than four . .	93
91.	Ovary 5-parted .	*The Limnanthe Tr.*
	Ovary undivided . . .	92
92.	Stamens distinct . .	*The Heath Tr.*
	Stamens monadelphous	*The Bread-tree Tr.*
93.	Calyx with two sepals .	*The Purslane Tr.*
	Calyx with more than two sepals .	94
94.	Stamens hypogynous . . .	95
	Stamens perigynous .	*The Celastrus Tr.*

95. Seeds with a tuft of hairs, or a hair *Tamarisk Tr.*
Seeds naked . *The Pittosporum Tr.*

96. Seeds numerous . *The Lythrum Tr.*
Seeds very few . *The Buckthorn Tr.*

97. Calyx altogether absent . . 98
Calyx present in some kind of state . 105

98. Leaves having stipules . . 99
Leaves destitute of stipules . . 102

99. Ovules very numerous . *The Willow Tr.*
Ovules very few . . . 100

100. Carp. triple; i. e. 3 consolidated *Euphorbia Tr.*
Carpels single . . . 101

101. Ovule erect; leaves fragrant . *The Gale Tr.*
Ovule pendulous; leaves scentless *The Plane Tr.*

102. Flowers unisexual . . . 103
Flowers bisexual . . *The Pepper Tr.*

103. Ovules naked; fruit in cones . *The Fir Tr.*
Ovules covered . . . 104

104. Carpels single . . *The Gale Tr.*
Carpels double . *The Callitriche Tr.*

105. Ovary more or less inferior . . 106
Ovary superior . . . 119

106. Leaves with stipules . . . 107
Leaves without stipules . . 109

107. Flowers bisexual . *The Birthwort Tr.*
Flowers unisexual . . . 108

108. Fruit in a cup, or cupule . *The Nut Tr.*
Fruit triangular, naked *The Begonia Tr.*

109. Flowers unisexual . . . 110
Flowers bisexual . . . 113

110. Flowers in catkins . . . 111
Flowers not in catkins . *The Gourd Tr.*

111. Leaves simple . . . 112
Leaves pinnated . . *The Walnut Tr.*

112. Leaves opposite . . *The Garrya Tr.*
Leaves alternate . . *The Gale Tr.*

113. Leaves with transparent dots . *The Myrtle Tr.*
Leaves dotless . . . 114

114. Ovary many-celled . . . 115
Ovary one-celled . . . 116

115. Ovary three or six-celled . *The Birthwort Tr.*
Ovary four-celled *The Evening Primrose Tr.*

116. Anther many-celled . *The Miselto Tr.*
Anther two-celled . . . 117

117. Stamens numerous, long . *The Combretum Tr.*
Stamens few and short . . 118

118. Embryo straight . *The Evening Primrose Tr.*
Embryo curved . *The Goosefoot Tr.*

119. Leaves with stipules . . . 120
Leaves without stipules . . 133

120. Flowers unisexual; that is, having stamens in one flower, and pistils in another 121
Flowers bisexual; that is, having stamens and pistils united in the same flower 123

121. Carpels solitary . *The Nettle Tr.*
Carpels more than one . . 122

122. Flowers in catkins . *The Birch Tr.*
Flowers not in catkins *The Euphorbia Tr.*

123. Sepals two . . *The Purslane Tr.*
Sepals more than two . . 124

124. Carpels solitary, or quite separate . 125
Carpels more than one, consolidated . 128

125. Fruit a legume . . *The Pea Tr.*
Fruit not a legume . . . 126

126. Calyx membranous . *The Knotgrass Tr.*
Calyx firm and herbaceous . . 127

127. One style to each ovary . *The Rose Tr.*
Three styles to each ovary *The Buckwheat Tr.*

128. Placentas parietal . *The Passion Flower Tr.*
Placentas in the axis . . . 129

129. Calyx membranous and ragged *The Elm Tr.*
Calyx firm and equally lobed . . 130

130. Calyx valvate . . . 131
Calyx imbricated . *The Geranium Tr.*

131. Stamens monadelphous . *The Sterculia Tr.*
Stamens distinct . . . 132

132. Stam. 4-5, opposite the petals *The Buckthorn Tr.*
Stamens 8-10 . . *The Linden Tr.*

133. Flowers bisexual; that is, having both stamens and pistil in the same flower 134
Flowers unisexual; that is, having stamens and pistils in separate flowers . 155

134. Sepals two . . *The Purslane Tr.*
Sepals more than two . . 135

135. Carpels several, consolidated . . 136
Carpels solitary, or if several quite distinct 144

136. Placentas parietal . . *The Poppy Tr.*
Placentas in the axis . . . 137

137. Number of ovules very small . . 138
Number of ovules very great . . 140

138. Leaves dotted . . *The Rue Tr.*
Leaves not dotted . . . 139

139. Embryo curved . *The Virginian Poke Tr.*
Embryo straight . *The Celastrus Tr.*

140. Carpels divaricating at point *The Saxifrage Tr.*
Carpels consolidated at the point . 141

141. Stamens perigynous . . . 142
Stamens hypogynons . . . 143

142. Fruit one-celled . . *The Primrose Tr.*
Fruit with several cells . *The Lythrum Tr.*

143. Calyx tubular . . *The Catchfly Tr.*
Calyx of distinct sepals *The Chickweed Tr.*

144. Carpels several . *The Crowfoot Tr.*
Carpels solitary . . . 145

145. Anther-valves recurved . . 146
Anther-valves straight . . 147

146. Leafy, erect, shrubs or trees *The Cinnamon Tr.*
Leafless, twining herbs . *The Cassytha Tr.*

147. Fruit a legume . *The Pea Tr.*
Fruit not a legume . . 148

148. Calyx hardened in the fruit . 149
Calyx always membranous . . 150

149. Base of calyx hardened *The Marvel of Peru Tr.*
Whole tube of calyx hardened *The Knawel Tr.*

150. Fruit triangular . *The Buckwheat Tr.*
Fruit round . . . 151

151. Stam. in the points of the sepals *The Protea Tr.*
Stamens not in the points of the sepals 152

152. Leaves covered with scurfiness *The Oleaster Tr.*
Leaves not scurfy . . . 153

153. Cal. tubular. Ovule pendulous *The Mezereum Tr.*
Calyx open and short. Ovule erect 154

154. Calyx dry and coloured *The Amaranth Tr.*
Calyx herbaceous . *The Goosefoot Tr.*

155. Stamens united in a column *The Pitcher-plant Tr.*
Stamens distinct . . . 156

156. Leaves dotted . *The Yellowwood Tr.*
Leaves not dotted . *The Euph. Tr.*

157. Ovary superior . . . 158
Ovary inferior . . . 194

158. Flowers regular . . . 159
Flowers irregular . . . 185

159. Ovary lobed . . . 160
Ovary not lobed . . . 162

160. Leaves dotted . . *The Rue Tr.*
Leaves dotless . . . 161

161. Flower-branches coiled up before opening *The Borage Tr.*
Flower-branches always straight *The Nolana Tr.*

162\. Anthers opening by pores . 163
Anthers opening by slits . . 166

163\. Carpels four or five . . 164
Carpels two . *The Nightshade Tr.*

164\. Herbaceous plants *The Winter-green Tr.*
Shrubs 165

165\. Anthers two-celled . *The Heath Tr.*
Anthers one-celled . *The Epacris Tr.*

166\. Carpels four or five. . . 167
Carpels three . . . 172
Carpels two . . . 174
Carpels one . . . 183

167\. Stamens opposite petals, and equal to them in number . . . 168
Stamens alternate with petals, or at least twice their number . . 169

168\. Herbaceous plants . *The Primrose Tr.*
Shrubs or trees . *The Ardisia Tr.*

169\. Brown parasites on roots *The Monotropa Tr.*
Leafy green plants . . 170

170\. Seeds very numerous . *The Houseleek Tr.*
Seeds very few . . . 171

171\. Ovules erect . *The Bindweed Tr.*
Ovules pendulous . *The Holly Tr.*

172\. Inflorescence coiled up . *The Hydrolea Tr.*
Inflorescence straight . . 173

173\. Anth. bursting longitud. *The Greek Valerian Tr.*
Anthers bursting transversely *The Diapensia Tr.*

174. Stamens two . . . 175
Stamens four or more . . 176

175. Æstivation of corolla valvate *The Olive Tr.*
Æstivation of corolla imbricated *The Jasmine Tr.*

176. Inflorescence coiled up *The Waterleaf Tr.*
Inflorescence straight . . 177

177. Æstivation of corolla plaited . 178
Æstivation of corolla imbricated . 180

178. Seeds very few . *The Bindweed Tr.*
Seeds very numerous . . 179

179. Leaves three-ribbed . *The Gentian Tr.*
Leaves one-ribbed . *The Nightshade Tr.*

180. Anthers adhering to a stigma-like table *The Asclepias Tr.*
Anthers quite free . . 181

181. Parasitical leafless plants . *The Dodder Tr.*
Green leafy terrestrial plants . . 182

182. Leaves uniformly three-ribbed *The Gentian Tr.*
Leaves one-ribbed . *The Wormseed Tr.*

183. Stigm. with an external covering *The Brunonia Tr.*
Stigma in its ordinary naked state . 184

184. Style one . . *The Plantain Tr.*
Styles five . . *The Leadwort Tr.*

185. Ovary four-lobed . . *The Mint Tr.*
Ovary undivided . . . 186

186. Carpel solitary . *The Madwort Tr.*
Carpels two 187

187. Fruit nut-like 188
Fruit capsular or succulent . . 189

188. Flowers without bracts . *The Myoporum Tr.*
Flowers with bracts . *The Vervain Tr.*

189. Seeds winged. Woody climbers
The Trumpet-flower Tr.
Seeds wingless . . . 190

190. Brown parasites . *The Broom Rape Tr.*
Green leafy plants . . . 191

191. Fruit two-celled . . . 192
Fruit with free centr. placenta *The Butterwort Tr.*

192. Ovary partly inferior . *The Gesnera Tr.*
Ovary quite superior . . 193

193. Seeds without appendages . *The Figwort Tr.*
Seeds with hooked appendages *The Justicia Tr.*

194. Carpel solitary . . . 195
Carpels more than one . . 197

195. Anthers grown together . *The Composite Tr.*
Anthers distinct . . . 196

196. Carpel quite solitary . *The Scabious Tr.*
Carpel with two additional abortive ones
The Valerian Tr.

197. Anthers grown together . *The Lobelia Tr.*
Anthers distinct . . . 198

198. Anthers opening by pores . *The Bilberry Tr.*
Anthers opening by slits . . 199

199. Stipules between opposite leaves *The Coffee Tr.*
Stipules absent . . . 200

200. Stigm. with an external covering *The Goodenia Tr.*
Stigma naked . . . 201

201. Style and stamens united in an irritable column . . *The Stylidium Tr.*
Style and stamens distinct . . 202

202. Seeds very numerous . *The Harebell Tr.*
Seeds very few . . . 203

203. Leaves alternate . . *The Ebony Tr.*
Leaves opposite . . . 204

204. Leaves in pairs. Stem round *The Honeysuckle Tr.*
Leaves in whorls. Stem square *The Madder Tr.*

205. Flowers incomplete; that is, not having distinct petals . . . 206
Flowers complete; that is, having distinct petals . . . 213

206. Flowers glumaceous . . . 207
Flowers not glumaceous . . 208

207. Stems round and hollow . *The Grass Tr.*
Stems solid . . . *The Sedge Tr.*

208. Flowers on a spadix . . . 209
Flowers scattered . . . 211

209. Fruit succulent . . *The Arum Tr.*
Fruit dry 210

210. Anthers sessile . . *The Acorus Tr.*
Anthers on long weak stalks . *The Bulrush Tr.*

211. Floaters 212
Land plants . *The Arrow-grass Tr.*

212. Ovules pendulous *The Naiad Tr.*
Ovules erect . . *The Duckweed Tr.*

213. Stamens and styles united in a central column . . *The Orchis Tr.*
Stamens and styles separate . . 214

214. Ovary inferior . . . 215
Ovary superior . . . 219

215. Veins of leaves diverging from midrib 216
Veins of leaves parallel with midrib . 217

216. Anther two-celled . *The Ginger Tr.*
Anther one-celled *The Arrow Root Tr.*

217. Stamens three . *The Cornflag Tr.*
Stamens six . . . 218
Stamens more than six . *The Frogbit Tr.*

218. Sepals thin and coloured *The Narcissus Tr.*
Sepals herbaceous . *The Pine-Apple Tr.*

219. Carpels quite separate . . 220
Carpels quite united . . 221

220. Fruits many-seeded *The Flowering Rush Tr.*
Fruits one-seeded *The Water-Plantain Tr.*

221. Sepals herbaceous; petals coloured *Spiderwort Tr.*
Sepals and petals both alike . 222

222. Flowers brown and glumaceous *The Rush Tr.*
Flowers coloured . . 223

223. Anthers turned outwards *The Colchicum Tr.*
Anthers turned inwards . . 224

224. Petals rolled inwards after flowering *Pontedera Tr.*
Petals shrivelling irregularly after flowering *The Lily Tr.*

225. Stems jointed . . . 225*
Stems not jointed . . 226

225* Fructification in cones *The Horsetail Tr.*
Fructification axillary and solitary *The Chara Tr.*

226. Plants with distinct leaves . 227
Plants mere leafless expansions . 230

227. Fructification growing on the back of the leaves . . *The Fern Tr.*
Fructification distinct from the leaves 228

228. Seed-vessel sessile in the axils of leaves *The Club-moss Tr.*
Seed-vessel on stalks . . 229

229. Seed-vessel with a lid and calyptra *The Moss Tr.*
Seed-vessel without lid and calyptra *The Jungermannia Tr.*

230. Seed-vessel opening into valves *Jungermannia Tr.*
Seed-vessel without valves . 231

231. Seed-vessel stalked and external *Marchantia Tr.*
Seed-vessel stalkless and usually internal 232

232. Growing under water *The Sea-weed Tr.*
Growing in the air . . 233

233. Fructification in external shields *The Lichen Tr.*
Fructification in internal cases *The Mushroom Tr.*

APPENDIX.

An Alphabetical List of the commoner kinds of Plants, with the Natural Orders to which they severally belong.

	Tribe or Natural Order.
ABELE Tree	Willow
Abies	Fir
Abrotanum	Composite
Abrus	Pea
Absinthium	Composite
Abutilon	Mallow
Acacia	Pea
Acanthus	Justicia
Acer	Maple
Aceras	Orchis
Achillea	Composite
Achyranthes	Amaranth
Aconitum	Crowfoot
Acorus	Acorus
Acrostichum	Fern
Actæa	Crowfoot
Acynos	Mint
Adam's Needle	Lily
Adder's tongue	Fern
Adenandra	Rue
Adenophora	Harebell
Adhatoda	Justicia

	Tribe or Natural Order.
Adiantum	Fern
Adlumia	Fumitory
Adonis	Crowfoot
Adoxa	Aralia
Ægilops	Grass
Ægopodium	Umbelliferous
Aerides	Orchis
Æsculus	Horsechesnut
Æthionema	Cruciferous
Æthusa	Umbelliferous
African marigold	Composite
Agapanthus	Lily
Agave	Narcissus
Ageratum	Composite
Agrimonia	Rose
Agrostemma	Lychnis
Air plant	Orchis
Ajuga	Mint
Alaternus	Buckthorn
Albuca	Asphodel
Alcea	Mallow
Alchemilla	Burnet

	Tribe or Natural Order.
Alder	Birch
Aletris	Asphodel
Alexanders	Umbelliferous
Alexandrian laurel	Asphodel
Alisma	Water Plantain
Alligator pear	Cinnamon
Allium	Asphodel
Allspice tree	Myrtle
Alnus	Birch
Aloe	Asphodel
Alonsoa	Foxglove
Alopecurus	Grass
Aloysia	Vervain
Alpinia	Ginger
Alsine	Chickweed
Alstrœmeria	Narcissus
Althæa	Mallow
Alyssum	Cruciferous
Amaryllis	Narcissus
Amelanchier	Apple
American aloe	Narcissus
American cowslip	Primrose
Ammobium	Composite
Amomum	Ginger
Amorpha	Pea
Ampelopsis	Vine
Amsonia	Asclepias
Amygdalus	Almond
Anagallis	Primrose
Anagyris	Pea
Ananassa	Pine Apple
Anchusa	Borage
Andersonia	Epacris
Andromeda	Heath
Andropogon	Grass
Androsæmum	Apocynum
Anemone	Crowfoot
Anethum	Umbelliferous
Angelica	ditto
Angelonia	Foxglove
Angræcum	Orchis
Anomatheca	Cornflag
Anthemis	Composite
Anthericum	Asphodel
Antholyza	Cornflag
Anthriscus	Umbelliferous
Anthyllis	Pea
Antirrhinum	Foxglove
Aotus	Pea
Aphelandra	Justicia
Apios	Pea
Apium	Umbelliferous
Aponogeton	Saururus
Apricot	Almond
Aquilegia	Crowfoot
Arabis	Cruciferous
Arachis	Pea
Araucaria	Fir
Arbor Vitæ	do.
Arbutus	Heath
Archangel	Umbelliferous
Archangelica	ditto
Arctium	Composite
Arctotis	ditto
Ardisia	Myrsine
Areca	Palm
Arenaria	Chickweed
Aretia	Primrose
Argemone	Poppy
Aristolochia	Birthwort

	Tribe or Natural Order.
Armeria	Thrift
Artemisia	Composite
Arthropodium	Asphodel
Artichoke	Composite
Artocarpus	Nettle
Arundo	Grass
Asarum	Birthwort
Asparagus	Asphodel
Asperula	Madder
Aspidium	Fern
Asplenium	do.
Aster	Composite
Astragalus	Pea
Astrantia	Umbelliferous
Astrapæa	Mallow
Astroloma	Epacris
Athamanta	Umbelliferous
Atragene	Crowfoot
Atriplex	Goosefoot
Atropa	Nightshade
Aubrietia	Cruciferous
Aucuba	Dogwood
Auricula	Primrose
Azalea	Heath
Azarolus	Apple
Babiana	Cornflag
Baccharis	Composite
Ballota	Mint
Balm	do.
Balm of Gilead	do.
Bamboo	Grass
Banksia	Protea
Baptisia	Pea
Barbarea	Cruciferous
Barleria	Justicia
Barley	Grass
Barringtonia	Myrtle
Bartonia	Loasa
Bartsia	Foxglove
Baryosma	Rue
Basil	Mint
Batatas	Bindweed
Batschia	Borage
Bauhinia	Pea
Bay tree	Cinnamon
Bean	Pea
Beaufortia	Myrtle
Beaumontia	Apocynum
Beccabunga	Foxglove
Beckmannia	Grass
Beech	Oak
Beet	Goosefoot
Belladonna Lily	Narcissus
Bellis	Composite
Bellium	ditto
Berberis	Barberry
Beta	Goosefoot
Betonica	Mint
Betula	Birch
Bidens	Composite
Bignonia	Trumpet Flower
Billardiera	Pittosporum
Billbergia	Pine Apple
Bird-cherry	Almond
Bird-pepper	Nightshade
Bird's-foot Trefoil	Pea
Biscutella	Cruciferous
Biserrula	Pea
Bitter-sweet	Nightshade
Bladder Ketmia	Mallow

	Tribe or Natural Order.
Bladder Senna	Pea
Blechnum	Fern
Bletia	Orchis
Blitum	Goosefoot
Blue-bottle	Composite
Blumenbachia	Loasa
Bocconia	Poppy
Bœhmeria	Nettle
Boltonia	Composite
Bombax	Mallow
Bonapartea	Pine Apple
Bonus Henricus	Goosefoot
Boronia	Rue
Botrychium	Fern
Bouvardia	Coffee
Box Tree	Spurge
Brachysema	Pea
Brachystelma	Asclepias
Brake	Fern
Bramble	Rose
Brasavola	Orchis
Brassia	ditto
Brassica	Cruciferous
Broccoli	ditto
Brodiæa	Asphodel
Bromelia	Pine Apple
Bromus	Grass
Brook-lime	Foxglove
Broom	Pea
Broughtonia	Orchis
Broussonetia	Nettle
Browallia	Foxglove
Brownea	Pea
Brugmansia	Nightshade
Brunsfelsia	ditto

	Tribe or Natural Order.
Brunsvigia	Narcissus
Bryonia	Gourd
Bryophyllum	Houseleek
Buddlea	Foxglove
Bugle	Mint
Bugloss	Borage
Bulbocodium	Colchicum
Bullace tree	Almond
Bupthalmum	Composite
Bupleurum	Umbelliferous
Burdock	Composite
Bur Reed	Bulrush
Butcher's Broom	Asphodel
Butomus	Flowering Rush
Butterfly plant	Asclepias
Butterwort	Crowfoot
Buxus	Spurge
Cabbage	Cruciferous
Cacalia	Composite
Cæsalpinia	Pea
Caladium	Arum
Calamintha	Mint
Calandrinia	Purslane
Calanthe	Orchis
Calathea	Arrow root
Calceolaria	Foxglove
Calendula	Composite
Callicarpa	Vervain
Calliopsis	Composite
Callistachys	Pea
Callistemon	Myrtle
Calluna	Heath
Calochortus	Lily
Calostemma	Narcissus
Calothamnus	Myrtle

	Tribe or Natural Order.
Calotropis	Asclepias
Caltha	Crowfoot
Calycanthus	Carolina Allspice
Calystegia	Bindweed
Camaridium	Orchis
Camellia	Tea
Cammarum	Crowfoot
Campanula	Harebell
Canarina	ditto
Canavalia	Pea
Candollea	Dillenia
Candytuft	Cruciferous
Canna	Arrow-root
Cannabis	Nettle
Canterbury Bells	Harebell
Capparis	Caper
Capraria	Foxglove
Caprifolium	Honeysuckle
Capsella	Cruciferous
Capsicum	Nightshade
Caragana	Pea
Caralluma	Asclepias
Caraway	Umbelliferous
Cardamine	Cruciferous
Cardoon	Composite
Carduus	ditto
Carex	Sedge
Carnation	Lychnis
Carob tree	Pea
Carrot	Umbelliferous
Carthamus	Composite
Caryophyllus	Myrtle
Cassia	Pea
Castanea	Oak
Castilleja	Foxglove

	Tribe or Natural Order.
Caster-oil plant	Spurge
Catananche	Composite
Catchfly	Lychnis
Catmint	Mint
Cat Thyme	ditto
Cattleya	Orchis
Caucalis	Umbelliferous
Cauliflower	Cruciferous
Ceanothus	Buckthorn
Cedar of Lebanon	Fir
Cedar	ditto
Celandine	Poppy
Celery	Umbelliferous
Celosia	Amaranth
Centaurea	Composite
Centaurium	Gentian
Cephalanthus	Coffee
Cerastium	Chickweed
Cerasus	Almond
Ceratonia	Pea
Cerbera	Apocynum
Cercis	Pea
Cereus	Cactus
Cerinthe	Borage
Ceropegia	Asclepias
Ceterach	Fern
Chærophyllum	Umbelliferous
Chamomile	Composite
Charlock	Cruciferous
Cheiranthus	ditto
Chelidonium	Poppy
Chelone	Foxglove
Chenopodium	Goosefoot
Cherimoyer	Anona
Cherry	Almond

	Tribe or Natural Order.
Chervil	Umbelliferous
Chimaphila	Winter-green
Chimonanthus	Carolina Allspice
China Aster	Composite
Chionanthus	Olive
Chironia	Gentian
Chives	Asphodel
Chlora	Gentian
Chorizema	Pea
Christmas Rose	Crowfoot
Christ's Thorn	Buckthorn
Chrysanthemum	Composite
Chrysosplenium	Saxifrage
Cicer	Pea
Cichorium	Composite
Cicuta	Umbelliferous
Cimicifuga	Crowfoot
Cineraria	Composite
Circæa,	Enchanter's Nightshade
Cissus	Vine
Cistus	Rock Rose
Citron	Orange
Citrus	ditto
Cladanthus	Composite
Clarkia	Evening Primrose
Clary	Mint
Claytonia	Purslane
Clematis	Crowfoot
Cleome	Caper
Clerodendrum	Vervain
Clethra	Heath
Clianthus	Pea
Cliffortia	Burnet
Clinopodium	Mint
Clintonia	Lobelia

	Tribe or Natural Order.
Clitoria	Pea
Clover	ditto
Cobæa	Greek Valerian
Coburghia	Narcissus
Coccoloba	Buckwheat
Cochineal Fig	Cactus
Cochlearia	Cruciferous
Cock's-comb	Amaranth
Cocos	Palm
Colletia	Buckthorn
Collinsia	Foxglove
Collinsonia	Mint
Collomia	Greek Valerian
Colutea	Pea
Comarum	Rose
Commelina	Spiderwort
Comptonia	Gale
Conferva	Sea weed
Conium	Umbelliferous
Convallaria	Asphodel
Convolvulus	Bindweed
Coptis	Crowfoot
Corallorrhiza	Orchis
Corchorus	Linden
Coriander	Umbelliferous
Cork tree	Oak
Corn-cockle	Lychnis
Cornus	Dogwood
Coronilla	Pea
Coronopus	Cruciferous
Corræa	Rue
Cortusa	Primrose
Corydalis	Fumitory
Corylus	Oak
Costmary	Composite

	Tribe or Natural Order.
Cotoneaster	Apple
Cotula	Composite
Cotyledon	Houseleek
Cow Parsley	Umbelliferous
Cowslip	Primrose
Crambe	Cruciferous
Crassula	Houseleek
Cratægus	Apple
Crinum	Narcissus
Crithmum	Umbelliferous
Crocus	Cornflag
Crotalaria	Pea
Croton	Spurge
Crowea	Rue
Cucubalus	Lychnis
Cucumber	Gourd
Cucumis	ditto
Cucurbita	ditto
Cunninghamia	Fir
Cuphea	Lythrum
Cupressus	Fir
Curcuma	Ginger
Currant	Gooseberry
Cuscuta	Dodder
Cyclamen	Primrose
Cydonia	Apple
Cymbidium	Orchis
Cynanchum	Asclepias
Cynara	Composite
Cynoglossum	Borage
Cynosurus	Grass
Cyperus	Sedge
Cypripedium	Orchis
Cyrtanthus	Narcissus
Cyrtopodium	Orchis
Cytisus	Pea
Daffodil	Narcissus
Dahlia	Composite
Daisy	ditto
Dalbergia	Pea
Dalibarda	Rose
Damasonium	Water Plantain
Dammar	Fir
Dandelion	Composite
Daphne	Mezereum
Darwinia	Pea
Datura	Nightshade
Daucus	Umbelliferous
Davallia	Fern
Daviesia	Pea
Deadly Nightshade	Nightshade
Dead Nettle	Mint
Delphinium	Crowfoot
Dendrobium	Orchis
Dens Canis	Lily
Dentaria	Cruciferous
Desmodium	Pea
Dianthus	Lychnis
Dictamnus	Rue
Diervilla	Honeysuckle
Digitalis	Foxglove
Dillwynia	Pea
Dioscorea	Yam
Diosma	Rue
Dipsacus	Scabious
Disa	Orchis
Disandra	Foxglove
Dodecatheon	Primrose
Dolichos	Pea
Doronicum	Composite

	Tribe or Natural Order.
Doryanthes	Narcissus
Dorycnium	Pea
Draba	Cruciferous
Dracæna	Asphodel
Dracocephalum	Mint
Dracontium	Arum
Drimia	Asphodel
Drosera	Sundew
Dryandra	Protea
Dryas	Rose
Duranta	Vervain
Eccremocarpus	Trumpet Flower
Echeveria	Houseleek
Echinocactus	Cactus
Echinops	Composite
Echites	Apocynum
Echium	Borage
Edwardsia	Pea
Elæagnus	Oleaster
Elichrysum	Composite
Elsholtzia	Poppy
Empetrum	Crowberry
Endive	Composite
English Mercury	Euphorbia
Epidendrum	Orchis
Epigæa	Heath
Epilobium	Evening Primrose
Epimedium	Berberry
Epiphyllum	Cactus
Equisetum	Horsetail
Eranthemum	Justicia
Eria	Orchis
Erica	Heath
Eriobotrya	Apple
Eriophorum	Sedge

	Tribe or Natural Order.
Eriostemon	Rue
Erodium	Geranium
Eruca	Cruciferous
Ervum	Pea
Eryngium	Umbelliferous
Erysimum	Cruciferous
Erythræa	Gentian
Erythrina	Pea
Erythronium	Lily
Eschscholtzia	Poppy
Esculus	Horsechesnut
Eucalyptus	Myrtle
Eucomis	Asphodel
Eugenia	Myrtle
Eulophia	Orchis
Euonymus	Celastrus
Eupatorium	Composite
Euphrasia	Foxglove
Eutoca	Waterleaf
Fagus	Oak
Farsetia	Cruciferous
Feather Grass	Grass
Fennel	Umbelliferous
Ferraria	Cornflag
Ferula	Umbelliferous
Feverfew	Composite
Ficaria	Crowfoot
Ficus	Nettle
Fig Tree	ditto
Filbert	Oak
Fontanesia	Olive
Fool's Parsley	Umbelliferous
Fragaria	Rose
Fraxinus	Olive
French Marigold	Composite

	Tribe or Natural Order.
Fritillary	Lily
Fuchsia	Evening Primrose
Fucus	Sea-weed
Fumaria	Fumitory
Furze	Pea
Gagea	Lily
Galanthus	Narcissus
Galardia	Composite
Galega	Pea
Galeobdolon	Mint
Galeopsis	ditto
Galium	Madder
Gardenia	Coffee
Gardoquia	Mint
Gaultheria	Heath
Gaura	Evening Primrose
Genista	Pea
Gerardia	Foxglove
Germander	Mint
Gethyllis	Narcissus
Geum	Rose
Gilia	Greek Valerian
Gillyflower	Cruciferous
Gladiolus	Cornflag
Glaucium	Poppy
Glaux	Primrose
Glechoma	Mint
Gleditschia	Pea
Globe Amaranth	Amaranth
Globe Thistle	Composite
Gloxinia	Gesnera
Glycine	Pea
Glycyrrhiza	ditto
Gnaphalium	Composite
Gnidia	Mezereum
Goat's Beard	Composite
Golden Rod	ditto
Golden Saxifrage	Saxifrage
Gomphrena	Amaranth
Gongora	Orchis
Goodia	Pea
Gordonia	Tea
Gorteria	Composite
Gossypium	Mallow
Grape	Vine
Grape Hyacinth	Asphodel
Gratiola	Foxglove
Grevillea	Protea
Grewia	Linden
Griffinia	Narcissus
Grindelia	Composite
Grislea	Lythrum
Ground Ivy	Mint
Groundsel	Composite
Guava	Myrtle
Guelder Rose	Honeysuckle
Guernsey Lily	Narcissus
Guilandina	Pea
Gum Cistus	Rock Rose
Gymnocladus	Pea
Gypsophila	Chickweed
Habenaria	Orchis
Habranthus	Narcissus
Hæmanthus	ditto
Hakea	Protea
Halesia	Styrax
Halimodendron	Pea
Hamamelis	Witch Hazel
Hawkweed	Composite
Hawthorn	Apple

	Tribe or Natural Order.
Hazel	Oak
Heart's Ease	Violet
Hedera	Aralia
Hedge Hyssop	Mint
Hedge Mustard	Cruciferous
Hedychium	Ginger
Hedysarum	Pea
Helenium	Composite
Helianthemum	Rock Rose
Helianthus	Composite
Helichrysum	ditto
Heliotropium	Borage
Hellebore	Crowfoot
Helonias	Colchicum
Hemerocallis	Lily
Hemimeris	Foxglove
Hemlock	Umbelliferous
Hemlock Spruce	Fir
Hemp	Nettle
Henbane	Nightshade
Hepatica	Crowfoot
Heracleum	Umbelliferous
Hermannia	Mallow
Hesperis	Cruciferous
Heuchera	Saxifrage
Hibbertia	Dillenia
Hibiscus	Mallow
Hieracium	Composite
Hippocrepis	Pea
Hippophae	Oleaster
Hippuris	Evening Primrose
Holly	Holly
Hollyhock	Mallow
Honesty	Cruciferous
Hop	Nettle
Horehound	Mint
Hornbeam	Oak
Horned Poppy	Poppy
Horseradish	Cruciferous
Hosackia	Pea
Hottonia	Primrose
Houseleek	Stonecrop
Houstonia	Coffee
Hovea	Pea
Hovenia	Buckthorn
Hoya	Asclepias
Humea	Composite
Humulus	Nettle
Hutchinsia	Cruciferous
Hyacinthus	Asphodel
Hydrocotyle	Umbelliferous
Hydrophyllum	Waterleaf
Hyoscyamus	Nightshade
Hypecoum	Poppy
Hypericum	Tutsan
Hypnum	Moss
Hypochæris	Composite
Hypoxis	Narcissus
Hyssop	Mint
Iberis	Cruciferous
Ilex	Holly
Illecebrum	Knotgrass
Impatiens	Balsam
Imperatoria	Umbelliferous
Indian Fig	Cactus
Indian Corn	Grass
Indian Shot	Arrow Root
Indigofera	Pea
Inga	ditto
Inula	Composite

	Tribe or Natural Order.
Ipomœa	Bindweed
Ipomopsis	Greek Valerian
Iris	Cornflag
Isatis	Cruciferous
Isopogon	Protea
Itea	Heath
Iva	Composite
Ivy	Aralia
Ixia	Cornflag
Ixora	Coffee
Jacaranda	Trumpet Flower
Jacobea Lily	Narcissus
Jambosa	Myrtle
Jasione	Lobelia
Jatropha	Spurge
Jerusalem Artichoke	Composite
Jonquil	Narcissus
Judas Tree	Pea
Juglans	Walnut
Jujube	Buckthorn
Julibrissin	Pea
Juniper	Fir
Kæmpferia	Ginger
Kalmia	Heath
Kaulfussia	Composite
Kennedia	Pea
Kerria	Rose
Kidney-bean	Pea
Kitaibelia	Mallow
Knautia	Scabious
Kölreuteria	Soapberry
Laburnum	Pea
Lachenalia	Asphodel
Lachnæa	Mezereum
Lactuca	Composite

	Tribe or Natural Order.
Ladanum	Rock Rose
Ladies' Slipper	Orchis
Lagerstrœmia	Lythrum
Lambertia	Protea
Lamb's Lettuce	Valerian
Lamium	Mint
Lantana	Vervain
Lapeyrousia	Asphodel
Lapsana	Composite
Larix	Iris
Larkspur	Crowfoot
Larochea	Stonecrop
Laserpitium	Umbelliferous
Lasiopetalum	Sterculia
Lathyrus	Pea
Laurestinus	Honeysuckle
Laurus	Cinnamon
Lavandula	Mint
Lavatera	Mallow
Lavender	Mint
Leschenaultia	Goodenia
Ledum	Heath
Lemon	Orange
Leontice	Barberry
Leontodon	Composite
Leonurus	Mint
Lepidium	Cruciferous
Leptosiphon	Greek Valerian
Limnocharis	Flowering Rush
Linaria	Foxglove
Linnæa	Honeysuckle
Linum	Flax
Liparia	Pea
Liparis	Orchis
Liriodendron	Magnolia

	Tribe or Natural Order.
Lithospermum	Borage
Littorella	Ribgrass
Loddigesia	Pea
Lomaria	Fern
Lomatia	Protea
London Pride	Saxifrage
Lonicera	Honeysuckle
Lopezia	Evening Primrose
Loquat	Apple
Lotus	Pea
Lousewort	Foxglove
Lovage	Umbelliferous
Love Apple	Nightshade
Love-lies-bleeding	Amaranth
Lucern	Pea
Lungwort	Borage
Lupine	Pea
Lycium	Nighsshade
Lycopodium	Clubmoss
Lycopsis	Borage
Lycopus	Mint
Lysimachia	Primrose
Macleaya	Poppy
Mahernia	Mallow
Mahonia	Barberry
Malcomia	Cruciferous
Malope	Mallow
Malva	ditto
Malvaviscus	ditto
Mammillaria	Cactus
Mandrake	Nightshade
Manettia	Coffee
Maranta	Arrow-root
Marica	Cornflag
Marigold	Composite

	Tribe or Natural Order.
Marjoram	Mint
Marrubium	ditto
Marsh Mallow	Mallow
Maurandya	Foxglove
Maxillaria	Orchis
Mays	Grass
Medicago	Pea
Medlar	Apple
Melaleuca	Myrtle
Melhania	Mallow
Melianthus	Rue
Melilotus	Pea
Melissa	Mint
Melittis	ditto
Melon	Gourd
Melon Thistle	Cactus
Mentha	Mint
Menyanthes	Gentian
Menziesia	Heath
Mercurialis	Spurge
Mesembryanthemum	Fig Marig.
Mespilus	Apple
Meum	Umbelliferous
Milfoil	Composite
Mimosa	Pea
Mimulus	Foxglove
Mirabilis	Marvel of Peru
Mitella	Saxifrage
Momordica	Gourd
Monanthes	Stonecrop
Monarda	Mint
Moræa	Cornflag
Moricandia	Cruciferous
Morus	Nettle
Mountain Ash	Apple

	Tribe or Natural Order.
Mucuna	Pea
Murraya	Orange
Murucuia	Passion Flower
Musa	Plantain
Muscari	Asphodel
Mustard	Cruciferous
Myosotis	Borage
Myrica	Gale
Myriophyllum	Evening Primrose
Myristica	Nutmeg
Myrrhis	Umbelliferous
Narthecium	Rush
Nasturtium	Cruciferous
Navelwort	Borage
Nectarine	Almond
Negundium	Maple
Nemophila	Waterleaf
Neottia	Orchis
Nepeta	Mint
Nerine	Narcissus
Nerium	Apocynum
N. Zealand Spinach	Tetragonia
Nicotiana	Nightshade
Nigella	Crowfoot
Nolana	Nolana
Nolitangere	Balsam
Nonea	Borage
Norfolk Island Pine	Fir
Norway Spruce	Fir
Nuphar	Water Lily
Nux-vomica	Apocynum
Nycterium	Nightshade
Nymphæa	Water Lily
Ocymum	Mint
Œnanthe	Umbelliferous

	Tribe or Natural Order.
Œnothera	Evening Primrose
Olea	Olive
Oleander	Apocynum
Oncidium	Orchis
Onion	Asphodel
Onobrychis	Pea
Ononis	ditto
Onopordum	Composite
Onosma	Borage
Ophioglossum	Fern
Ophrys	Orchis
Opuntia	Cactus
Orach	Goosefoot
Origanum	Mint
Ornithogalum	Asphodel
Ornithopus	Pea
Ornus	Olive
Orobus	Pea
Orontium	Arum
Osbeckia	Melastoma
Osier	Willow
Osmunda	Fern
Ostrya	Oak
Othonna	Composite
Oxalis	Woodsorrel
Oxycoccus	Bilberry
Oxylobium	Pea
Pachysandra	Spurge
Pæonia	Crowfoot
Paliurus	Buckthorn
Palma Christi	Spurge
Panax	Aralia
Pancratium	Narcissus
Papaver	Poppy
Pardanthus	Cornflag

	Tribe or Natural Order.
Parietaria	Nettle
Paris	Colchicum
Parkinsonia	Pea
Parnassia	Saxifrage
Paronychia	Knotgrass
Parsley	Umbelliferous
Parsnep	ditto
Passerina	Mezereum
Pastinaca	Umbelliferous
Patersonia	Cornflag
Patrinia	Valerian
Pavetta	Coffee
Pavonia	Mallow
Peach	Almond
Pear	Apple
Pedicularis	Foxglove
Pelargonium	Geranium
Peltaria	Cruciferous
Pennyroyal	Mint
Pentapetes	Mallow
Penthorum	Stonecrop
Pentstemon	Foxglove
Peperomia	Pepper
Peppermint	Mint
Pereskia	Cactus
Pergularia	Asclepias
Periploca	ditto
Periwinkle	Apocynum
Persoonia	Protea
Petunia	Nightshade
Phaca	Pea
Phacelia	Waterleaf
Phalangium	Asphodel
Pharnaceum	Chickweed
Phaseolus	Pea
Pheasant's Eye	Crowfoot
Phillyrea	Olive
Phleum	Grass
Phlomis	Mint
Phlox	Greek Valerian
Phormium	Lily
Photinia	Apple
Phycella	Narcissus
Phylica	Buckthorn
Phyllanthus	Spurge
Physalis	Nightshade
Pimelea	Mezereum
Pimenta	Myrtle
Pimpernel	Primrose
Pimpinella	Umbelliferous
Pinaster	Fir
Pinckneya	Coffee
Pine	Fir
Pink	Lychnis
Pinus	Fir
Pitcairnia	Pine Apple
Planera	Elm
Plantago	Ribgrass
Platanus	Plane
Plumbago	Leadwort
Plumieria	Apocynum
Podalyria	Pea
Podophyllum	Poppy
Poinciana	Pea
Polemonium	Greek Valerian
Polyanthes	Lily
Polyanthus	Primrose
Polygala	Milkwort
Polygonum	Buckwheat
Polypodium	Fern

	Tribe or Natural Order.		Tribe or Natural Order.
Pomaderris .	Buckthorn	Rampion .	Harebell
Pomegranate .	Myrtle	Ranunculus .	Crowfoot
Poplar . .	Willow	Rape .	Cruciferous
Populus . .	ditto	Raphanus .	ditto
Portugal Laurel	Almond	Raphiolepis .	Apple
Portulaca .	Purslane	Rapistrum	Cruciferous
Potato .	Nightshade	Raspberry . .	Rose
Potentilla . .	Rose	Red Cedar . .	Fir
Poterium .	Burnet	Renanthera .	Orchis
Pothos . .	Arum	Renealmia .	Ginger
Pot Marigold	Composite	Reseda .	Mignonette
Prince's Feather	Amaranth	Restharrow .	Pea
Prinos . .	Celastrus	Rhamnus .	Buckthorn
Privet . .	Olive	Rheum .	Buckwheat
Prunella .	Mint	Rhexia .	Melastoma
Prunus . .	Almond	Rhinanthus .	Foxglove
Psidium . .	Myrtle	Rhipsalis .	Cactus
Psoralea . .	Pea	Rhodiola .	Stonecrop
Pteris . .	Fern	Rhododendron .	Heath
Pulmonaria .	Borage	Rhodora .	ditto
Pulsatilla .	Crowfoot	Rhubarb .	Buckwheat
Pultenæa . .	Pea	Rhus .	Cashew
Pumpkin . .	Gourd	Ribes .	Gooseberry
Punica . .	Myrtle	Ricinus . .	Spurge
Puschkinia .	Asphodel	Robinia . .	Pea
Pyracantha .	Apple	Rose Acacia .	ditto
Pyrola .	Wintergreen	Rose Campion .	Lychnis
Pyrus . .	Apple	Rosemary . .	Mint
Quercus . .	Oak	Rosmarinus .	ditto
Quince . .	Apple	Rubia . .	Madder
Quisqualis	Combretum	Rubus . .	Rose
Radiola . .	Flax	Rudbeckia .	Composite
Radish .	Cruciferous	Rumex .	Buckwheat
Ragged Robin	Lychnis	Ruscus .	Asphodel
Ragwort .	Composite	Russelia .	Foxglove

	Tribe or Natural Order.
Ruta	Rue
Saccharum	Grass
Saffron	Cornflag
Sage	Mint
Sagina	Chickweed
Sagittaria	Water Plantain
Saintfoin	Pea
St. John's Bread	ditto
St. John's Wort	Tutsan
Salicornia	Goosefoot
Salisburia	Fir
Salix	Willow
Salpiglossis	Foxglove
Salsafy	Composite
Salsola	Goosefoot
Salvia	Mint
Sambucus	Honeysuckle
Samolus	Primrose
Samphire	Umbelliferous
Sanguinaria	Poppy
Sanguisorba	Burnet
Sanicula	Umbelliferous
Sanseviera	Asphodel
Santolina	Composite
Saponaria	Lychnis
Satureja	Mint
Satyrium	Orchis
Savin	Fir
Savory	Mint
Scandix	Umbelliferous
Schizanthus	Foxglove
Schizopetalon	Cruciferous
Schœnus	Sedge
Scilla	Asphodel
Scleranthus	Knawel
Scolopendrium	Fern
Scolymus	Composite
Scoparia	Foxglove
Scorpiurus	Pea
Scorzonera	Composite
Scrophularia	Foxglove
Scurvy Grass	Cruciferous
Scutellaria	Mint
Sea Buckthorn	Oleaster
Sea Kale	Cruciferous
Sedum	Stonecrop
Sempervivum	ditto
Senecio	Composite
Sensitive Plant	Pea
Service	Apple
Sesleria	Grass
Shaddock	Orange
Shallot	Asphodel
Shepherdia	Oleaster
Sibbaldia	Rose
Siberian Crab	Apple
Sibthorpia	Foxglove
Sicyos	Gourd
Sida	Mallow
Sideritis	Mint
Sieversia	Rose
Silene	Lychnis
Silphium	Composite
Sinapis	Cruciferous
Sisymbrium	ditto
Sisyrinchium	Cornflag
Sium	Umbelliferous
Smilacina	Asphodel
Smyrnium	Umbelliferous
Snapdragon	Foxglove

	Tribe or Natural Order.		Tribe or Natural Order.
Snowball Tree	Honeysuckle	Stevia .	Composite
Snowberry .	ditto	Stipa . .	Grass
Snowdrop .	Narcissus	Stock .	Cruciferous
Snowdrop Tree .	Styrax	Stork's Bill	Geranium
Solanum .	Nightshade	Stramonium .	Nightshade
Soldanella .	Primrose	Strawberry .	Rose
Solidago .	Composite	Strawberry Blite	Goosefoot
Sonchus . .	ditto	Strawberry Tree .	Heath
Sophora . .	Pea	Strelitzia .	Plantain
Sorrel .	Buckwheat	Struthiòla .	Mezereum
Southernwood	Composite	Stuartia .	Tea
Sowthistle .	ditto	Styphelia .	Epacris
Sparaxis .	Cornflag	Subularia .	Cruciferous
Sparganium	Bulrush	Succory .	Composite
Sparrmannia .	Linden	Sugar Cane .	Grass
Spartium . .	Pea	Sunflower .	Composite
Spergula .	Chickweed	Sutherlandia .	Pea
Sphacele .	Mint	Swainsona .	ditto
Spinach .	Goosefoot	Sweet Bay .	Cinnamon
Spindle tree .	Celastrus	Sweet Briar .	Rose
Spiræa . .	Rose	Sweet Flag .	Acorus
Sprengelia .	Epacris	Sweet Gale .	Gale
Spurge Laurel	Mezereum	Sweet Marjoram .	Mint
Squill .	Asphodel	Sweet Pea .	Pea
Squirting Cucumber	Gourd	Sweet Sultan .	Composite
Stachys . .	Mint	Sweet William	Lychnis
Stachytarpheta .	Vervain	Sycamore .	Maple
Stanhopea .	Orchis	Symphoria	Honeysuckle
Stapelia .	Asclepias	Symphytum .	Borage
Staphylea .	Bladder-nut	Syringa . .	Olive
Star of Bethlehem	Asphodel	Tacsonia	Passion-flower
Star of the Earth	Cruciferous	Tagetes .	Composite
Statice .	Leadwort	Talinum .	Purslane
Stellaria .	Chickweed	Tamarindus	Pea
Sternbergia .	Narcissus	Tamarix .	Tamarisk

	Tribe or Natural Order.
Tamus	Yam
Tanacetum	Composite
Tare	Pea
Taxodium	Fir
Taxus	ditto
Teak Wood	Vervain
Teasel	Scabious
Tea Tree	Tea
Tecoma	Trumpet-flower
Tectona	Vervain
Telephium	Stonecrop
Tellima	Saxifrage
Telopea	Protea
Templetonia	Pea
Terminalia	Combretum
Tetragonolobus	Pea
Teucrium	Mint
Thalictrum	Crowfoot
Thea	Tea
Theophrasta	Myrsine
Thermopsis	Pea
Thistle	Composite
Thlaspi	Cruciferous
Thorn Apple	Nightshade
Thrift	Leadwort
Thrincia	Composite
Thuja	Fir
Thunbergia	Justicia
Thyme	Mint
Tiarella	Saxifrage
Tigridia	Cornflag
Tilia	Linden
Tillandsia	Pine Apple
Tofieldia	Colchicum
Tormentilla	Rose

	Tribe or Natural Order.
Trachelium	Harebell
Tradescantia	Spiderwort
Tragopogon	Composite
Trapa	Evening Primrose
Traveller's Joy	Crowfoot
Trefoil	Pea
Tree Onion	Asphodel
Trifolium	Pea
Trigonella	ditto
Trillium	Colchicum
Triteleia	Asphodel
Triticum	Grass
Tritoma	Asphodel
Trollius	Crowfoot
Tropæolum	Nasturtium
Tulip	Lily
Tulip Tree	Magnolia
Turnip	Cruciferous
Turritis	ditto
Tussilago	Composite
Typha	Bulrush
Ulex	Pea
Ulmus	Elm
Urtica	Nettle
Vaccinium	Bilberry
Valantia	Madder
Vallota	Narcissus
Vanda	Orchis
Vanguiera	Coffee
Vanilla	Vanilla
Vella	Cruciferous
Veltheimia	Asphodel
Venus' Fly-trap	Sundew
Venus' Looking-glass	Harebell
Veratrum	Colchicum

	Tribe or Natural Order.
Verbascum	Foxglove
Verbena	Vervain
Veronica	Foxglove
Vesicaria	Cruciferous
Vestia	Nightshade
Vetch	Pea
Viburnum	Honeysuckle
Vicia	Pea
Vinca	Apocynum
Virgilia	Pea
Virginian Creeper	Vine
Virgin's Bower	Crowfoot
Viscum	Misseltoe
Vitex	Vervain
Vitis	Vine
Wallflower	Cruciferous
Water Cress	ditto
Water Hemlock	Umbelliferous
Water Horehound	Mint
Water Melon	Gourd
Watsonia	Cornflag
Wayfaring-tree	Honeysuckle
Welsh Onion	Asphodel
Westringia	Mint
Wheat	Grass
White Cedar	Fir
White Clover	Pea
White Spruce	Fir
Whortle Berry	Bilberry
Winter Aconite	Crowfoot
Winter Cherry	Nightshade
Witheringia	ditto
Woodbine	Honeysuckle
Wood Sage	Mint
Woodsia	Fern
Woodwardia	ditto
Wormwood	Composite
Wulfenia	Foxglove
Xanthorhiza	Crowfoot
Xeranthemum	Composite
Xerophyllum	Colchicum
Xerotes	Rush
Yew Tree	Fir
Yucca	Lily
Zea	Grass
Zebra Plant	Arrow Root
Zephyranthes	Narcissus
Zieria	Rue
Zingiber	Ginger
Zinnia	Composite
Zizyphus	Buckthorn

INDEX

OF THE SECOND VOLUME.

THE NUMBERS REFER TO THE PAGES.

THE END.

NORMAN AND SKEEN, PRINTERS, MAIDEN LANE, COVENT GARDEN

Printed in the United States
By Bookmasters